THE SCIENCE OF DISCWORLD IV

WHERE'S MY COW? (illustrated by Melvyn Grant)

THE ART OF DISCWORLD (with Paul Kidby)

THE WIT AND WISDOM OF DISCWORLD (compiled by Stephen Briggs)

THE FOLKLORE OF DISCWORLD (with Jacqueline Simpson)

MISS FELICITY BEEDLE'S THE WORLD OF POO
(assisted by Bernard and Isobel Pearson)

─────────── **Discworld Maps and Gazetteers** ───────────

THE STREETS OF ANKH-MORPORK
(with Stephen Briggs, painted by Stephen Player)

THE DISCWORLD MAPP (with Stephen Briggs, painted by Stephen Player)

A TOURIST GUIDE TO LANCRE – A DISCWORLD MAPP
(with Stephen Briggs, illustrated by Paul Kidby)

DEATH'S DOMAIN (with Paul Kidby)

THE COMPLEAT ANKH-MORPORK (with the Discworld Emporium)

A complete list of Terry Pratchett ebooks and audio books as well as
other books based on the Discworld series – illustrated screenplays,
graphic novels, comics and plays – can be found on
www.terrypratchett.co.uk

─────────── **Non-Discworld books** ───────────

THE DARK SIDE OF THE SUN • STRATA

THE UNADULTERATED CAT (illustrated by Gray Jolliffe)

GOOD OMENS (with Neil Gaiman)

THE LONG EARTH (with Stephen Baxter)

A BLINK OF THE SCREEN: COLLECTED SHORT FICTION

─────────── **Non-Discworld novels for younger readers** ───────────

THE CARPET PEOPLE • TRUCKERS • DIGGERS • WINGS

ONLY YOU CAN SAVE MANKIND • JOHNNY AND THE DEAD

JOHNNY AND THE BOMB • NATION • DODGER

THE SCIENCE OF
DISCWORLD IV
JUDGEMENT DAY

Terry Pratchett
Ian Stewart & Jack Cohen

EBURY
PRESS

Imagination bodies forth
The forms of things unknown, the poet's pen
Turns them to shapes and gives to airy nothing
A local habitation and a name.

WILLIAM SHAKESPEARE
A MIDSUMMER NIGHT'S DREAM

We don't see things as they are.
We see them as we are.

ANAÏS NIN

Philosophy is questions that may never be answered.
Religion is answers that may never be questioned.

ANONYMOUS, FROM DANIEL DENNETT
BREAKING THE SPELL: RELIGION AS A NATURAL PHENOMENON

Even scientists believe in God.
They've found Him in the Large Hadron Kaleidoscope.

DOOR-TO-DOOR MISSIONARY
REPORTED IN *NEW SCIENTIST*

'I never made the world,' said Om. 'Why should I make the world?
It was here already. And if I *did* make a world, I wouldn't make it
a ball. People'd fall off. All the sea'd run off the bottom.'

SMALL GODS

'Come on into the Library. It's got an earthed copper roof,
you know. Gods really hate that sort of thing.'

SMALL GODS

CONTENTS

PROLOGUE

WORLDS, DISC AND ROUND

 There is a sensible way to make a world.

It should be flat, so that no one falls off acciden-tally* unless they get too near the edge, in which case it's their own fault.

It should be circular, so that it can revolve sedately to create the slow progression of the seasons.

It should have strong supports, so that it doesn't fall down.

The supports should rest on firm foundations.

To avoid an infinite regression, the foundations should do what foundations are supposed to do, and stay up of their own accord.

It should have a sun, to provide light. This sun should be small and not too hot, to save energy, and it should revolve around the disc to separate day from night.

The world should be populated by people, since there is no point in making it if no one is going to live there.

Everything should happen because people want it to (magic) or because the power of story (narrativium) demands it.

This sensible world is Discworld – flat, circular, held up by four world-bearing elephants standing firmly on the back of a giant space-faring turtle and inhabited by ordinary humans, wizards,

---

* Falling off deliberately is another matter, about which they can be as imaginative as they wish. See *The Light Fantastic*, *The Colour of Magic* and *The Last Hero*.

witches, trolls, dwarves, vampires, golems, elves, the tooth fairy and the Hogfather.

But—

There is also a stupid way to make a world. And sometimes, that is necessary.

When an experiment in fundamental thaumaturgy on the squash court of Unseen University ran wild and threatened to destroy the universe, the computer HEX had to use up a huge quantity of magic in an instant. The only option was to activate the Roundworld Project, a magical force field that – paradoxically – keeps magic out. When the Dean of Unseen University poked his finger in to see what would happen, Roundworld switched on.

Roundworld isn't entirely sure which bit of itself its name applies to. Sometimes the name refers to the planet, sometimes to the entire universe. There have been a few mishaps along the way, but the Roundworld universe has now been running fairly successfully for thirteen and a half billion years; all of it started by an old man with a beard.

In the absence of magic, and lacking natural narrativium, the Roundworld universe runs on rules. Not rules made by people, but rules made by Roundworld itself; which is weird, because Roundworld has no idea what its rules ought to be. It seems to make them up as it goes along, but it's hard to be sure.

Certainly, it doesn't know what size it ought to be. From outside, as it gathers dust on a shelf in Rincewind's office, the Roundworld universe – a globe about 20 centimetres in diameter – resembles a cross between a foot-the-ball and a child's snowstorm toy. From inside, it appears to be somewhat larger: a sphere whose radius is about 400 sextillion kilometres. As far as its only known* inhabitants can tell, it may be much larger still; perhaps even infinite.

Such a huge universe seems to be cosmic overkill, because those inhabitants occupy only the tiniest part of its awe-inspiring volume,

---

* This may be misleading since it is the opinion of the inhabitants concerned.

namely the surface of an approximate sphere a mere twelve thousand kilometres across.

The wizards call this sphere Roundworld too. Its inhabitants call it Earth, because that's what the surface is usually made of (except for the wet, rocky, sandy and icy bits): a typically parochial attitude. Until a few centuries ago they thought that Earth was fixed at the centre of the universe; the rest, which revolved around it or wandered crazily across the sky, was of minor importance since it didn't contain *them*.

Roundworld the planet, as the name suggests, is *round*. Not round like a disc, but round like a foot-the-ball. It is younger than Roundworld the universe: about one third of its age. Though cosmically minuscule, the planet is fairly big compared to its inhabitants, so that if you live there, and you're stupid, you can be fooled into imagining that it's flat.

To prevent the planet's inhabitants falling off, the rules state that a mysterious force glues them on. Thankfully, there are no world-bearing elephants. If there were, the inhabitants would be able to walk *round* their world to the point where it meets an elephant. This world-bearing beast of immense power would appear to be *lying on its back*, its feet in the air. (Paint the soles yellow and you wouldn't be able to see it floating in a bowl of custard …)

Roundworld's rules are democratic. Not only does this mysterious force glue people to their world: it glues everything to everything else. But the glue is weak, and everything can – and usually does – move.

This includes Roundworld the planet. It does have a sun, but this sun does not go round the planet. Instead, *the planet goes round the sun*. Worse, that doesn't create day and night; instead, it produces seasons, because the planet is tilted. Also, the orbit isn't circular. It's a bit squashed, which is typical of Roundworld's jerry-built construction. So to get day and night, the planet has to spin as well. It works, in its way: if you're really stupid, you can be fooled into imagining that the sun goes round the planet. But – wouldn't you just know it – the spin also prevented Roundworld from being a sensible sphere, because when it was molten it got sort of squashed, just like its orbit … oh, forget it.

As a consequence of this hopelessly bungled arrangement, the sun has to be enormous, and a very long distance away. So it has to be ridiculously hot: so hot that special new rules have to come into play to allow it to burn. And then almost all of its prodigious energy output is wasted, trying to warm up empty space.

Roundworld has no supports. It appears to think it's a turtle, because it swims through space, tugged along by those mysterious forces. Its human inhabitants are not bothered by a sphere that swims, despite the absence of flippers. But then, people turned up at most four hundred thousand years ago, one hundredth of a per cent of the lifetime of the planet. And they seem to have turned up by accident, starting out as little blobs and then spontaneously becoming more complex – but they argue a lot about that. They're not terribly bright, to be honest, and they only started to work out modern scientific rules of the universe they live in four hundred years ago, so they've got a lot of catching up to do.

The inhabitants refer to themselves optimistically as *Homo sapiens*, meaning 'wise man' in an appropriately dead language. Their activities seldom fit that description, but there are occasional glorious exceptions. They should really be called *Pan narrans*, the storytelling ape, because nothing appeals to them more than a rollicking good yarn. They are narrativium incarnate, and they are currently refashioning their world to resemble Discworld, so that things *do* happen because people want them to. They have invented their own form of magic, with spells like 'make a dugout canoe', 'switch on the light', and 'login to Twitter'. This kind of magic cheats by using the rules behind the scenes, but if you're really, really stupid you can ignore that and pretend it's magic.

The first *The Science of Discworld* explained all that, and much more, including the giant limpet and the ill-fated crab civilisation's great leap sideways. An endless series of natural disasters established something that the wizards intuitively knew from the word go: a round world is not a safe place to be. Fast-forwarding through Roundworld history, they managed to skip from some not very promising apes huddled around a black monolith to the collapse of the space elevators, as some presumably highly intelligent creatures, having finally got the message, fled the planet and headed for the stars to escape yet another ice age.

They couldn't really be descended from those apes, could they? The apes seemed to have only two interests: sex, and bashing each other over the head.

In *The Science of Discworld II*, the wizards were surprised to find that the intelligent star-farers were indeed descended from the apes – a strange new use of the word 'descend', and one that caused serious trouble later. They found that out because Roundworld had taken the wrong leg of the Trousers of Time and had therefore deviated from its original timeline. Its ape-derived humans had become barbarians, their society vicious and riddled with superstition. They would never leave the planet in time to escape their doom. Something had interfered with Roundworld's history.

Feeling somehow responsible for the planet's fate, much as one might worry about a sick gerbil, the wizards entered their bizarre creation, to find that it was infested by elves. Discworld's elves are not the noble creatures of some Roundworld myths. If an elf told you to eat your own head, you'd do it. But going back in time to when the elves had arrived, and kicking them out, just made everything worse. The evil had gone, but it had taken with it any shred of innovation.

Examining Roundworld's history on what ought to have been its correct timeline, the wizards deduced that two key people – prominent among those very few wise ones – had never been born. This omission had to be repaired to get the planet back on track. They were William Shakespeare, whose artistic creations would give birth to a genuine spirit of humanity, and Isaac Newton, who would provide science. With considerable difficulty, and some interesting failures along the

way requiring ceilings to be painted black, the wizards nudged humanity back onto the only timeline that would save it from annihilation.
Shakespeare's *A Midsummer Night's Dream* tipped the tables decisively
by exposing the elves to ridicule. Newton's *Principia Mathematica*
completed the job by pointing humanity at the stars. Job done.

It couldn't last.

By the time of *The Science of Discworld III*, Roundworld was in
trouble again. Having safely entered its Victorian era, which should
have been a hotbed of innovation, it had once more departed from
its proper history. New technology was developing, but at a snail's
pace. Some vital spur to innovation had been lost, and the gerbil of
humanity was sick once more. This time, a key figure had written
the wrong book. The Reverend Charles Darwin's *Theology of Species*,
explaining the complexity of life through divine intervention, had
been so well received that science and religious belief had converged.
The creative spark of rational debate* had been lost. By the time the
Reverend Richard Dawkins finally wrote *The Origin of Species (by
Means of Natural Selection &c &c &c …)* it was too late to develop
space travel before the ice came down.

This time, getting Darwin born was not the problem. Getting
him to write the correct book … that was where everything went
pear-shaped, and it proved remarkably hard to nudge history back
on track. Contrary to the proverb, supplying a missing nail from
a horse's shoe does not save a kingdom. It generally has no effect,
aside from making the horse feel a bit more comfortable, because
hardly anything important has a single cause. It took a huge squad
of wizards, making over two thousand carefully choreographed
changes, to get Darwin onto the *Beagle*, stop him jumping ship when
he was being as sick as a dog, and perk his interest in geology so that
he stayed with the expedition† until it got to the Galápagos Islands.

---

* That is, insults, name-calling and shameless point-scoring.
† Loosely speaking. He remained on land whenever feasible, about 70% of the entire
'voyage'.

They wouldn't have succeeded at all, but the wizards eventually realised that something was actively interfering with their efforts to reset history to manufacturer's specifications. The Auditors of Reality are the ultimate Health and Safety officers: they much prefer a universe in which nothing interesting ever happens, and they are willing to go to extreme lengths to ensure that it doesn't. They had been blocking the wizards' every move.

It was a near thing. Even when the wizards successfully arranged for Darwin to visit the Galápagos and notice the finches and mockingbirds and turtles, it took years for him to understand the significance of those creatures – by which time all the turtle shells were long gone, tossed overboard after their contents were eaten, and he'd given away the finches to a bird expert. (He *had* realised that the mockingbirds were interesting.) It took even longer to get him to take the plunge and write *The Origin* instead of *The Ology*; he kept writing scholarly books about barnacles instead. Then, when he had finally managed to write *The Origin*, he still messed up with *Origin II*, calling it *The Descent of Man* – oh dear. *The Ascent of Man* would have been a better marketing ploy.

Anyway, the wizards finally achieved success, even contriving to bring Darwin into Discworld to meet the God of Evolution and admire the wheels on his elephant. The publication of *The Origin* established the corresponding timeline as the only one that had ever happened. (The Trousers of Time are like that.) Roundworld was saved *again*, and could rest undisturbed on its shelf, gathering dust …

Until—

ONE

GREAT BIG THING

 Every university must have, sooner or later, a big or, more preferably, a *Great Big Thing*. According to Ponder Stibbons, head of Inadvisably Applied Magic at Unseen University, it was, he said, practically a law of nature; and it couldn't be too big, and it had to be a thing, and definitely not a small one.

The senior wizards, eyeing the chocolate biscuits on the tray brought in by the tea lady, listened with as much attention as could be expected from wizards momentarily afflicted with chocolate starvation. Ponder's carefully written and argued speech pointed out that studious research throughout Library-space, or L-space as it is colloquially known, revealed that not to have a Great Big Thing would be a pitiful thing; and the lack of such a thing, indeed, in the academic universe, would make the university they were sitting in right now the butt of jokes and sardonic jibes by people who would be ashamed to be called their fellow academics – said jibes being all the more painful because academics know what *sardonic* actually means.

And when Mister Stibbons finished his last well-tuned argument, Mustrum Ridcully, the Archchancellor, put his hand heavily on the last disputed chocolate biscuit and said, 'Well now, Ponder, if I know you, and I most certainly *do* know you, then you never put in front of me a problem without having a proposed solution somewhere up your sleeve.' Ridcully's eyes narrowed a little as he continued, 'Indeed, Mister Stibbons, it would be very unlike you not already to have a Great Big Candidate. Am I not right?'

Ponder didn't bother to blush, but simply said, 'Well, sir, I do know that we in the HEM* do think that there are many puzzles presented to us by the universe that we really need to solve. As they say, sir: what you don't know can kill you! Ha-ha.'

Ponder was pleased with coming up with that remark; he knew his Archchancellor – who had the instincts of a fighter, and a bare-knuckle fighter at that – and so he moved in with, 'I'm thinking of the fact that we simply don't know why there is a third slood derivative, which in theory means that at the birth of the universe, in that very first nanosecond, the universe actually began to travel backwards in time. According to Von Flamer's experiment, that means that we appear to be coming and going at the same time! Ha-ha!'

'Yes, well, I can quite believe that,' said Ridcully glumly, looking at his fellows; and because he was the Archchancellor, after all, he added, 'Wasn't there something about a cat that was alive and dead at the same time?'

Ponder was always ready for this sort of thing and he said, 'Yes, sir, but it was only a hypothetical cat, sir, as it turned out – nothing to get pet-owners all upset about – and may I add that the elastic string theory turned out to be just one more unproven hypothesis, as did the bubble theory of interconnecting horizons.'

'Really.' Ridcully sighed. 'What a shame. I rather liked that one. Oh well, I trust that in its short life it gave some theoretical scientists a living, and so happily its little life wasn't wasted. You know, Mister Stibbons, over the years you have often discoursed with me about the various theories, hypotheses, concepts and conjectures in the world of natural science. You know what? I just wonder, I really do wonder, whether the universe – being of course by its very nature, dynamic, and possibly in some curious sense *sapient* – may now perhaps be trying to escape from your incessant prying, and is possibly driving you into even greater feats of intellect. The little tease!'

---

* High Energy Magic department.

There was a pause from the assembled wizards, and for a moment the face of Ponder Stibbons appeared to be made of polished bronze; then he said, 'What an amazing deduction, Archchancellor. I applaud you! Everybody knows that Unseen University will rise to meet *any* challenge; with your permission, sir, I will set to work on a budget right now. The Roundworld project was only a beginning. Now, with the ... Challenger Project, we will explore the fundamental basis of magic in our world!'

He ran to the High Energy Magic building so fast that his progress metamorphosed into a hurtle, which in ballistic terms is exactly the opposite of a turtle and *extremely* more streamlined.

And *that* was six years ago ...

Today, Lord Vetinari, tyrant of Ankh-Morpork, glanced up at the Great Big Thing which appeared to be doing nothing but humming to itself. It hovered in the air, appearing and disappearing, and in Vetinari's opinion looking somewhat smug, a feat indeed for something that had no face.

It was, in fact, a rather amorphous blob that seemed to twist magical equations with arcane symbols and squiggles that clearly meant *something* to those who knew about such things. The Patrician was not, on his own admission, a lover of technical things that spun and, indeed, hummed. Nor of unidentifiable squiggles. He saw them as things with which you couldn't negotiate, or argue; you couldn't hang them either, or even creatively torture them. Of course, the dictum *noblesse oblige* came to the rescue as always – although those who knew Havelock Vetinari well knew that he sometimes wasn't all that obliging.

On this occasion Lord Vetinari was being introduced to excitable and occasionally spotty young wizards in white robes – though still of course in pointy hats – who made a great fuss about large conglomerations of mindless and humming machinery behind the blob. Nevertheless, he did his best to look enthusiastic, and managed to drum up some conversation with Mustrum Ridcully, the Archchancellor, who it seemed was just as much in the dark as himself; and he congratulated Ridcully because it was clearly the thing to do, whatever the thing *did*.

'I'm sure you must be very proud, Archchancellor. It's extremely good, clearly a triumph, most certainly!'

Ridcully chuckled and said, 'Bravo! Thank you *so* much, Havelock, and do you know what? *Some* people said that if we turned the experiment on it would bring the world to an end! Can you imagine that? Us! The psychic protectors of the city, and indeed of the world throughout history!'

Lord Vetinari took an almost imperceptible step back and carefully enquired, 'And precisely when was it that you *did* turn it on, may I ask? It seems to be humming along quite adequately at the moment.'

'As a matter of fact, Havelock, the humming is going to end very shortly. The noise you are hearing is coming from a swarm of bees in the garden over there, and the Bursar hasn't had enough time to instruct them to get back to work. In fact, we were hoping that you would do the honours after lunch, if it is all right by you, of course?'

The expression on the face of Lord Havelock Vetinari was, for a moment, a picture: and it was a picture painted by a *very* modern artist, one who had been smoking something generally considered to turn the brain to cheese.

But *noblesse oblige* was a crushing imperative even for a tyrant, especially one who valued his self-esteem, and therefore, two hours later, a well-fed Lord Vetinari stood in front of the huge humming thing, feeling rather concerned. He made a small oration on the need for mankind to further its knowledge of the universe.

'While it is still there,' he added, looking very pointedly at Ridcully.

Then, after posing for the iconographer's lenses, he looked at the big red button on its stand in front of him and thought, I wonder if there *is* any truth in the rumours that this could end the world? Well, it's too late now to protest, and it would be quite remiss for me to draw back at this point. He brightened up and thought, If indeed it's me who blows up the known world, then it might just be good for my image anyway.

He pressed the button to the kind of applause people make when they understand that something important has happened while at the same time having no idea what they are really cheering. After checking, Vetinari turned to the Archchancellor and said, 'It would seem, Mustrum, that I have not destroyed the universe, which is something of a comfort. Is anything else supposed to happen?'

The Archchancellor slapped him on the back and said, 'Don't fret, Havelock: the Challenger Project was started up yesterday evening by Mister Stibbons over a cup of tea, just to make certain that it would start; and seeing that it was warmed up, he left it on. This of course in no way demeans *your* part in the ceremony, I promise you. The formality of the *significant* opening is at the heart of the whole business, which I am proud to say has all gone swimmingly!'

And *that* was six minutes ago …

TWO

GREAT BIG THINKING

 Great Big Things have a seductive allure, to which Roundworld's scientists are by no means immune. Most science requires relatively modest equipment, some is inherently expensive, and some would finance a small nation. Governments worldwide are addicted to big science, and often find it easier to authorise a ten-billion dollar project than one costing ten thousand – much as a committee will agree to a new building in five minutes, but then spend an hour debating the cost of biscuits. We all know why: it takes an expert to evaluate the design and price of a building, but everyone understands biscuits. The funding of big science is sometimes depressingly similar. Moreover, for administrators and politicians seeking to enhance their careers, big science is more prestigious than small science, because it involves more money.

However, there can also be a more admirable motive for huge scientific projects: big problems sometimes require big answers. Putting together a faster-than-light drive on the kitchen table using old baked bean cans may work in a science fiction story, but it's seldom a realistic way to proceed. Sometimes you get what you pay for.

Big science can be traced back to the Manhattan project in World War II, which developed the atomic bomb. This was an extraordinarily complex task, involving tens of thousands of people with a variety of skills. It stretched the boundaries of science, engineering and, above all, organisation and logistics. We don't want to suggest that finding really effective ways to blow people to smithereens is necessarily a sensible criterion for success, but the Manhattan project convinced a lot of people that big science can have a huge impact on the entire planet. Governments have promoted big science ever since; the Apollo Moon landings and the human genome project are familiar examples.

Some areas of science are unable to function at all without Great Big Things. Perhaps the most prominent is particle physics, which has given the world a series of gigantic machines, called particle accelerators, which probe the small-scale structure of matter. The most powerful of these are colliders, which smash subatomic particles into stationary targets, or into each other in head-on collisions, to see what gets spat out. As particle physics progresses, the new particles that theorists are predicting become more exotic and harder to detect. It takes a more energetic collision to spit them out, and more mathematical detective work and more powerful computers to compile evidence that they were, for an almost infinitesimal moment of time, actually present. So each new accelerator has to be bigger, hence more expensive, than its predecessors.

The latest and greatest is the Large Hadron Collider (LHC). 'Collider' we know about, 'hadron' is the name of a class of subatomic particles, and 'large' is fully justified. The LHC is housed in two circular tunnels, deep underground; they are mostly in Switzerland but wander across the border into France as well. The main tunnel is eight kilometres across, and the other one is about half as big. The tunnels contain two tubes, along which the particles of interest – electrons, protons, positrons and so on – are propelled at speeds close to that of light by 1,624 magnets. The magnets have to be kept at a temperature close to absolute zero, which requires 96 tonnes of liquid helium; they are absolutely enormous, and most weigh over 27 tonnes.

The tubes cross at four locations, where the particles can be smashed into each other. This is the time-honoured way for physicists to probe the structure of matter, because the collisions generate a swarm of other particles, the bits and pieces out of which the original particles are made. Six enormously complex detectors, located at various points along the tunnels, collect data on this swarm, and powerful computers analyse the data to work out what's going on.

The LHC cost €7.5 billion – about £6 billion or $9 billion – to build. Not surprisingly, it is a multinational project, so big politics gets in on the act as well.

Ponder Stibbons has two reasons for wanting a Great Big Thing. One is the spirit of intellectual enquiry, the mental fuel on which the High Energy Magic building runs. The bright young wizards who inhabit that building want to discover the fundamental basis of magic, a quest that has led them to such esoteric theories as quantum thaumodynamics and the third slood derivative, as well as the fateful experiment in splitting the thaum that inadvertently brought Round-world into existence in the first place. The second reason opened the previous chapter: every university that wants to be *considered* a university has to have its very own Great Big Thing.

It is much the same in Roundworld – and not only for universities.

Particle physics began with small equipment and a big idea. The word 'atom' means 'indivisible', a choice of terminology that was a hostage to fortune from the day it was minted. Once physicists had swallowed the proposition that atoms exist, which they did just over a century ago, a few began to wonder if it might be a mistake to take the name literally. In 1897 Joseph John Thomson showed that they had a point when he discovered cathode rays, tiny particles emanating from atoms. These particles were named electrons.

You can hang around waiting for atoms to emit new particles, you can encourage them to do so, or you can make them an offer they can't refuse by bashing them into things to see what breaks off and where it goes. In 1932 John Cockroft and Ernest Walton built a small particle accelerator and memorably 'split the atom'. It soon emerged that atoms are made from three types of particle: electrons, protons and neutrons. These particles are extremely small, and even the most powerful microscopes yet invented cannot make them visible –

though atoms can now be 'seen' using very sensitive microscopes that exploit quantum effects.

All of the elements – hydrogen, helium, carbon, sulphur and so on – are made from these three particles. Their chemical properties differ because their atoms contain different numbers of particles. There are some basic rules. In particular, the particles have electrical charges: negative for the electron, positive for the proton, and zero for the neutron. So the number of protons should be the same as the number of electrons, to make the total charge zero. A hydrogen atom is the simplest possible, with one electron and one proton; helium has two electrons, two protons and two neutrons.

The main chemical properties of an atom depend on the number of electrons, so you can throw in different numbers of neutrons without changing the chemistry dramatically. However, it does change a bit. This explains the existence of isotopes: variants of a given element with subtly different chemistry. An atom of the commonest form of carbon, for instance, has six electrons, six protons and six neutrons. There are other isotopes, which have between two and sixteen neutrons. Carbon-14, used by archaeologists to date ancient organic materials, has eight neutrons. An atom of the commonest form of sulphur has sixteen electrons, sixteen protons and sixteen neutrons; 25 isotopes are known.

Electrons are especially important for the atom's chemical properties because they are on the outside, where they can make contact with other atoms to form molecules. The protons and neutrons are clustered closely together at the centre of the atom, forming its nucleus. In an early theory, electrons were thought to orbit the nucleus like planets going round the Sun. Then this image was replaced by one in which an electron is a fuzzy probability cloud, which tells us not where the particle *is*, but where it is *likely* to be found if you try to observe it. Today, even that image is seen as an oversimplification of some pretty advanced mathematics in which the electron is nowhere and everywhere at the same time.

Those three particles – electrons, protons and neutrons – unified the whole of physics and chemistry. They explained the entire list of chemical elements from hydrogen up to californium, the most complex naturally occurring element, and indeed various short-lived man-made elements of even greater complexity. To get matter in all its glorious variety, all you needed was a short list of particles, which were 'fundamental' in the sense that they couldn't be split into even smaller particles. It was simple and straightforward.

Of course, it didn't *stay* simple. First, quantum mechanics had to be introduced to explain a vast range of experimental observations about matter on its smallest scales. Then several other equally fundamental particles turned up, such as the photon – a particle of light – and the neutrino – an electrically neutral particle that interacts so rarely with everything else that it would be able to pass though thousands of miles of solid lead without difficulty. Every night, countless neutrinos generated by nuclear reactions in the Sun pass right through the solid Earth, and through you, and hardly any of them have any effect on anything.

Neutrinos and photons were only the beginning. Within a few years there were more fundamental particles than chemical elements, which was a bit worrying because the explanation was becoming more complicated than the things it was trying to explain. But eventually physicists worked out that some particles are more fundamental than others. A proton, for example, is made from three smaller particles called quarks. The same goes for the neutron, but the combination is different. Electrons, neutrinos and photons, however, remain fundamental; as far as we know, they're not made out of anything simpler.*

---

* Ever since the 1970s physicists have speculated that quarks and electrons are actually made from even smaller particles, variously named alphons, haplons, helons, maons, prequarks, primons, quinks, rishons, subquarks, tweedles and Y-particles. The generic name for such particles is currently 'preon'.

One of the main reasons for constructing the LHC was to investigate the final missing ingredient of the standard model, which despite its modest name seems to explain almost everything in particle physics. This model maintains, with strong supporting evidence, that *all* particles are made from sixteen truly fundamental ones. Six are called quarks, and they come in pairs with quirky names: up/down, charmed/strange, and top/bottom. A neutron is one up quark plus two down quarks; a proton is one down quark plus two up quarks.

Next come six so-called leptons, also in pairs: the electron, muon, and tauon (usually just called tau) and their associated neutrinos. The original neutrino is now called the electron neutrino, and it is paired with the electron. These twelve particles – quarks and leptons – are collectively called fermions, after the great Italian-born American physicist Enrico Fermi.

The remaining four particles are associated with forces, so they hold everything else together. Physicists recognise four basic forces of nature: gravity, electromagnetism, the strong nuclear force and the weak nuclear force. Gravity plays no role in the standard model because it hasn't yet been fitted into a quantum-mechanical picture. The other three forces are associated with specific particles known as bosons in honour of the Indian physicist Satyendra Nath Bose. The distinction between fermions and bosons is important: they have different statistical properties.

The four bosons 'mediate' the forces, much as two tennis players are held together by their mutual attention to the ball. The electromagnetic force is mediated by the photon, the weak nuclear force is mediated by the Z-boson and the W-boson, and the strong nuclear force is mediated by the gluon. So that's the standard model: twelve fermions (six quarks, six leptons) held together by four bosons.

Sixteen fundamental particles.

Oh, and the Higgs boson – *seventeen* fundamental particles.

Assuming, of course, that the fabled Higgs (as it is colloquially called) actually existed. Which, until 2012, was moot.

Despite its successes, the standard model fails to explain why most particles have masses (for one particular technical meaning of 'mass'). The Higgs came to prominence in the 1960s, when several physicists realised that a boson with unusual features might solve one important aspect of this riddle. Among them was Peter Higgs, who worked out some of the hypothetical particle's properties and predicted that it should exist. The Higgs boson creates a Higgs field: a sea of Higgs bosons. The main unusual feature is that the strength of the Higgs field is not zero, even in empty space. When a particle moves through this all-pervasive Higgs field it interacts with it, and the effect can be interpreted as mass. One analogy is moving a spoon through treacle, but that misrepresents mass as resistance, and Higgs is critical of that way of describing his theory. Another analogy views the Higgs as a celebrity at a party, who attracts a cluster of admirers.

The existence (or not) of the Higgs boson was the main reason, though by no means the only one, for spending billions of euros on the LHC. And in July 2012 it duly delivered, with the announcement by two independent experimental teams of the discovery of a previously unknown particle. It was a boson with a mass of about 126 GeV (billion electronvolts, a standard unit used in particle physics), and the observations were consistent with the Higgs in the sense that those features that could be measured were what Higgs had predicted.

This discovery of the long-sought Higgs, if it holds up, completes the standard model. It could not have been made without big science, and it represents a major triumph for the LHC. However, the main impact to date has been in theoretical physics. The existence of the Higgs does not greatly affect the rest of science, which already assumes that particles have mass. So it could be argued that the same amount of money, spent on less spectacular projects, would almost certainly have produced results with more practical utility. However, it is in the nature of Great Big Things that if the money isn't spent on *them*, it isn't spent on smaller scientific projects either.

Small projects don't advance bureaucratic or political careers as effectively as big ones.

The discovery of the Higgs exemplifies some basic issues about how scientists view the world, and about the nature of scientific knowledge. The actual evidence for the Higgs is a tiny bump on a statistical graph. In what sense can we be confident that the bump actually represents a new particle? The answer is extremely technical. It is impossible to observe a Higgs boson directly, because it splits spontaneously and very rapidly into a swarm of other particles. These collide with yet other particles, creating a huge mess. It takes very clever mathematics, and very fast computers, to tease out of this mess the characteristic signature of a Higgs boson. In order to be sure that what you've seen isn't just coincidence, you need to observe a large number of these Higgs-like events. Since they are very rare, you need to run the experiments many times and perform some sophisticated statistical analysis. Only when the chance of that bump being coincidence falls below one in a million do physicists allow themselves to express confidence that the Higgs is real.

We say 'the' Higgs, but there are alternative theories with more than one Higgs-like particle – *eighteen* fundamental particles. Or nineteen, or twenty. But now we know there is at least one, when before it might have been none.

Understanding all this requires considerable expertise in esoteric areas of theoretical physics and mathematics. Even understanding the aspect of 'mass' involved, and which particles it applies to, is complicated. Performing the experiment successfully requires a range of engineering skills, in addition to a deep background in experimental physics. Even the word 'particle' has a technical meaning, nothing like the comfortable image of a tiny ball bearing. So in what sense can scientists claim to 'know' how the universe behaves, on such a small scale that no human can perceive it directly? It's not like looking through a telescope and seeing that Jupiter has four smaller bodies going round it, as Galileo did; or like looking down a microscope and realising that living things are made from tiny cells, as Robert Hooke did. The evidence for the Higgs, like that for most basic aspects of science, is not exactly in your face.

To come to grips with these questions, we take a look at the nature of scientific knowledge, using more familiar examples than the Higgs. Then we distinguish two fundamentally different ways to think about the world, which will form a running theme throughout the book.

Science is often thought to be a collection of 'facts', which make unequivocal statements about the world. The Earth goes round the Sun. Prisms separate light into its component colours. If it quacks and waddles, it's a duck. Learn the facts, master the technical jargon (here being: orbit, spectrum, *Anatidae*), tick the boxes, and you understand science. Government administrators in charge of education often take this view, because they can count the ticks (*Ixodidae* – no, scratch that).

Oddly, the people who disagree most strongly are scientists. They know that science is nothing of the kind. There are no hard-and-fast facts. Every scientific statement is provisional. Politicians hate this. How can anyone trust scientists? If new evidence comes along, they change their minds.

Of course, some parts of science are less provisional than others. No scientist expects the accepted description of the shape of the Earth to change overnight from round to flat. But they have already seen it change from a plane to a sphere, from a sphere to a spheroid flattened at the poles, and from a perfect spheroid to a bumpy one. A recent press release announced that the Earth is shaped like a lumpy potato.* On the other hand, no one would be surprised if new measurements revealed that the Earth's seventeenth spherical harmonic – one component of the mathematical description of

---

* Provided all irregularities are exaggerated by a factor of 7000. http://www.new scientist.com/article/dn20335-earth-is-shaped-like-a-lumpy-potato.html

its shape – needed to be increased by two per cent. Most changes in science are gradual and progressive, and they don't affect the big picture.

Sometimes, however, the scientific worldview changes radically. Four elements became 98 (now 118 as we've learned how to make new ones). Newton's gravity, a force acting mysteriously at a distance, morphed into Einstein's curved spacetime. Fundamental particles such as the electron changed from tiny hard spheres to probability waves, and are now considered to be localised excitations in a quantum field. The field is a sea of particles and the particles are isolated waves in that sea. The Higgs field is an example: here the corresponding particles are Higgs bosons. You can't have one without the other: if you want to be a particle physicist, you have to understand the physics of quantum fields as well. So the word 'particle' necessarily acquires a different meaning.

Scientific revolutions don't change the universe. They change how humans interpret it. Many scientific controversies are mainly about interpretations, not 'the facts'. For example, many creationists don't dispute the *results* of DNA sequencing;* instead, they dispute the interpretation of those results as evidence for evolution.

Humans are hot on interpretation. It lets them wriggle out of awkward positions. In 2012, in a televised debate about sexism in religion and the vexed issue of female bishops in the Church of England, some months before the General Synod voted against the proposal, one participant quoted 1 Timothy 2:12-14: 'But I suffer not a woman to teach, nor to usurp authority over the man, but to be in silence. For Adam was first formed, then Eve. And Adam was not deceived, but the woman being deceived was in the transgression.' It

---

* Recall that DNA stands for 'deoxyribonucleic acid', a type of molecule that famously takes the form of a double helix, like two interwound spiral staircases. The 'steps' of the staircase come in four kinds, called bases, which are like code letters. The sequence of bases differs from one organism to the next, and it represents genetic information about that organism.

seems hard to interpret this as anything other than a statement that women are inferior to men, that they should be subservient and *shut up*, and that moreover, original sin is entirely the fault of women, not men, because Eve fell for the serpent's temptation. Despite this apparently unequivocal reading, another participant stoutly maintained that the verses meant nothing of the kind. It was just a matter of interpretation.

Interpretations matter, because 'the facts' seldom explain how the universe relates to *us*. 'The facts' tell us that the Sun's heat comes from nuclear reactions, mainly hydrogen fusing to helium. But we want more. We want to know *why*. Did the Sun come into existence *in order to* provide us with heat? Or is it the other way round: are we on this planet because the Sun's heat provided an environment in which creatures like us could evolve? The facts are the same either way, but their implications depend on how we interpret them.

Our default interpretation is to view the world in human terms. This is no great surprise. If a cat has a point of view, it surely views the world in feline terms. But humanity's natural mode of operation has had a profound effect on how we think about our world, and on what kinds of explanation we find convincing. It also has a profound effect on *what world we think about*. Our brains perceive the world on a human scale, and interpret those perceptions in terms of what is – or sometimes was – important *to us*.

Our focus on the human scale may seem entirely reasonable. How else would we view our world? But rhetorical questions deserve rhetorical answers, and for us, unlike the rest of the animal kingdom, there are alternatives. The human brain can consciously modify its own thought-patterns. We can teach ourselves to think on other scales, both smaller and larger. We can train ourselves to avoid psychological traps, such as believing what we want to *because* we want to. We can think in even more alien ways: mathematicians routinely contemplate spaces with more than three dimensions, shapes so complicated that they have no meaningful volume, surfaces with only one side, and different sizes of infinity.

Humans *can* think inhuman thoughts.

That kind of thinking is said to be analytic. It may not come naturally, and its outcomes may not always be terribly comforting, but it's *possible*. It has been the main path to today's world, in which analytic thinking has become increasingly necessary for our survival. If you spend your time comfortably telling yourself that the world is what you want it to be, you will get some nasty surprises, and it may be too late to do anything about them. Unfortunately, the need to think analytically places a huge barrier between science and many human desires and beliefs that re-emerge in every generation. Battles scientists fondly imagined were won in the nineteenth century must continually be re-fought; rationality and evidence alone may not be enough to prevail.

There is a reason for our natural thought-patterns. They evolved, along with us, because they had survival value. A million years ago, human ancestors roamed the African savannahs, and their lives depended – day in, day out – on finding enough food to keep them alive, and avoiding becoming food themselves. The most important things in their lives were their fellow human beings, the animals and plants that they ate, and the animals that wanted to eat *them*.

Their world also included many things that were not alive: rocks; rivers, lakes and seas; the weather; fires (perhaps started by lightning); the Sun, Moon and stars. But even these often seemed to share some of the features of life. Many of them moved; some changed without any apparent pattern, as if acting on their own impulses; and many could kill. So it is not surprising that as human culture developed, we came to view our world as the outcome of conscious actions by living entities. The Sun, Moon and stars were gods, visible evidence for the existence of supernatural beings that lived in the heavens. A rumble of thunder, a flash of lightning – these were signs of the gods' displeasure. The evidence was all around us on a daily basis, which put it beyond dispute.

In particular, animals and plants were central to the lives of early humans. You only have to browse through a book of Egyptian hieroglyphs to notice just how many of them are animals, birds, fish, plants ... or bits of animals, birds, fish and plants. Egyptian gods were depicted with the heads of animals; in one extreme case, the god Khepri, the head was an entire dung beetle, neatly placed on top of an otherwise headless human body. Khepri was one aspect of the Sun-god, and the dung beetle (or scarab) got in on the act because dung beetles roll balls of dung around and dig them into the ground. Therefore the Sun, a giant ball, is pushed around by a giant dung beetle; as proof, the Sun also disappears into the ground (the underworld) every evening at sunset.

The physicist and science fiction author Gregory Benford has written many essays with a common theme: broadly speaking, human styles of thought tend to fall into two categories.* One is to see humanity as the context for the universe; the other is to see the universe as the context for humanity. The same person can think both ways of course, but most of us tend to default to one of them. Most ways to separate people into two kinds are nonsense: as the old joke goes, there are two kinds of people: those who think there are two kinds of people, and those who don't. But Benford's distinction is an illuminating one, and it holds more than a grain of truth.

We can paraphrase it like this. Many people see the surrounding world – the universe – as a resource for humans to exploit; they also see it as a reflection of themselves. What matters most, in this view, is always human-centred. 'What can this do *for me?*' (or '*for us?*') is the main, and often the only, question worth asking. From such a viewpoint, to understand something is to express it in terms of human agency. What matters is its *purpose*, and that is whatever *we* use it

* Gregory Benford, a creature of double vision, in *Science Fiction and the Two Cultures: Essays on Bridging the Gap between the Sciences and the Humanities*, edited by Gary Westfahl and George Slusser, McFarland Publishers 2009, pages 228-236.

for. In this worldview, rain exists in order to make crops grow and to provide fresh water for us to drink. The Sun is there because it warms our bodies. The universe was designed with us in mind, constructed so that we could live in it, and it would have no meaning if we were not present.

It is a short and natural step to see human beings as the pinnacle of creation, rulers of the planet, masters of the universe. Moreover, you can do all of that without any conscious recognition of how narrowly human-centred your worldview is, and maintain that you are acting out of humility, not arrogance, because of course we are subservient to the universe's creator. Which is basically a superhuman version of us – a king, an emperor, a pharaoh, a lord – whose powers are expanded to the limits of our imagination.

The alternative view is that human beings are just one tiny feature of a vast cosmos, most of which does not function on a human scale or take any notice of what we want. Crops grow because rain exists, but rain exists for reasons that have virtually nothing to do with crops. Rain has been in existence for billions of years, crops for about ten thousand. In the cosmic scheme of things, human beings are just one tiny incidental detail on an insignificant ball of rock, most of whose history happened before we turned up to wonder what was going on. We may be the most important thing in the universe as far as we are concerned, but nothing that happens outside our tiny planet depends on our existence, with a few obvious exceptions like various small but complicated bits of metal and plastic now littering the surface of the Moon and Mars, in orbit around Mercury, Jupiter and Saturn, or wandering through the outer edges of our solar system. We might say that the universe is indifferent to us, but even that statement is too self-conscious; it endows the universe with the human attribute of indifference. There is no 'it' to be indifferent. The system of the world does not function in human terms.

We'll refer to these ways of thinking as 'human-centred' and 'universe-centred'. Many controversies that grab the headlines stem, to a greater or lesser extent, from the deep differences between them. Instead of assuming that one must be superior to the other, and then arguing vehemently about which one it is, we should first learn to recognise the difference. Both have advantages, in their proper spheres of influence. What causes trouble is when they tread on each other's toes.

Before the early twentieth century, scientists used to think that phenomena like light could either be particles or waves, but not both. They argued – often nastily – about which was correct. When quantum theory was invented, it turned out that matter had both aspects, inseparably intertwined. At about the time that all reputable scientists *knew* that light was a wave, photons turned up, and those were particles of light. Electrons, which were obviously particles when they were discovered, turned out to have wavelike features as well. So quantum physicists got used to the idea that things that seemed to be particles were actually tiny clumps of waves.

Then quantum field theory came along, and the waves stopped being clumped. They could spread out. So now particle physicists have to know about quantum fields, and our best explanation of why 'particles' have mass is the existence of an all-pervading Higgs field. On the other hand, the current evidence only supports the existence of the particle-like aspect of this field: the Higgs boson. The field itself has not been observed. It might not exist, and that would be interesting, because it would overturn the way physicists currently think about particles and fields. It would also be somewhat annoying.

In everyday life, we encounter solid, compact objects, such as rocks, and they make it easy for us to think about tiny particles. We encounter sloshy but well-defined structures that move around on water, and we feel comfortable with waves. In a human-centred view, there are no sloshy rocks, which makes us assume – almost without questioning it – that nothing can be both particle and wave at the same time. But universe-centred thinking has shown that this assumption can be wrong outside the human domain.

The human-centred view is as old as humanity itself. It seems to be the default pattern of thinking for most of us, and that makes sound evolutionary sense. The universe-centred view appeared more recently. In the sense that we're thinking of – science and the scientific method – universe-centred thinking has become widespread only in the last three or four hundred years. It is still a minority view, but a very influential one. To see why, we must understand two things: how science goes about its business, and what constitutes scientific evidence.

For those of us who are willing to pay attention, the universe-centred view has revealed just how big, how ancient, and how awe-inspiring the universe is. Even on a human scale, it's a very impressive place, but our parochial perceptions pale into insignificance when confronted by the mind-numbing reality.

When early humans roamed the plains of Africa, the world must have *seemed* huge, but it was actually extremely small. A big distance was what you could walk in a month. An individual's experience of the world was limited to the immediate region in which he or she lived. For most purposes, a human-centred view works very well for such a small world. The important plants and animals – the ones useful to specific groups of humans – were relatively few in number, and located in their immediate vicinity. One person could encompass them all, learn their names, know how to milk a goat or to make a roof from palm fronds. The deeper message of the Egyptian hieroglyphs is not how diverse that culture's flora and fauna were, but how narrowly its symbolism was tailored to the organisms that were important to everyday Egyptian life.

As we came to understand our world more deeply, and asked new questions, comfortable answers in terms that we could intuitively understand began to make less and less sense. Conceivably the Sun might, metaphorically, be pushed around by an invisible giant dung-beetle, but the Sun is a vast ball of very hot gas and no ordinary beetle could survive the heat. You either fix things up by attributing supernatural powers to your beetle, or you accept that a beetle can't hack it. You then have to accept that the motion of the Sun occurs for reasons that differ significantly from the purposeful shoving of a beetle storing up food for its larvae, raising the interesting question 'why or how *does* it move?'. Similarly, although the setting Sun looks as if it is disappearing underground, you can come to understand that it is being obscured by the rotating bulk of the Earth. Instead of telling a story that offers little real insight, you've learned something new about the world.

It took time for humanity to realise all this, because our planet is far larger than a village. If you walked 40 kilometres every day it would take you three years to travel all the way round the world, ignoring ocean crossings and other obstacles. The Moon is nearly ten times as far away; the Sun is 390 times as far away as the Moon. To get to the nearest star, you must multiply that figure by a further 270,000. The diameter of our home galaxy is 25,000 times as great again. The nearest galaxy of comparable size, the Andromeda galaxy, is 25 times as far away. The distance from Earth to the edge of the observable visible universe is more than 18,000 times as great as that. In round figures, 400,000,000,000,000,000,000,000 kilometres.

Four hundred sextillion. That's some village.

We have no intuitive feel for anything that large. In fact, we have little intuitive feel for distances of more than a few thousand miles, and those only because many of us now travel such distances by air – which shrinks the world to a size we can comprehend. From London, New York is just a meal away.

We know that the universe is that big, and that old, because we have developed a technique that consciously and deliberately sets aside the human-centred view of the world. It does so by searching not just for evidence to confirm our ideas, which human beings have done since the dawn of time, but for evidence that could disprove them, a new and rather disturbing thought. This technique is called science. It replaces blind faith by carefully targeted doubt. It has existed in its current form for no more than a few centuries, although precursors go back a few thousand years. There is a sense in which 'know' is too strong a word, for scientists consider all knowledge to be provisional. But what we 'know' through science rests on much more secure foundations than anything else that we claim to know, because those foundations have survived being tested to destruction.

Through science, we know how big and how old the Earth is. We know how big and how old our solar system is. We know how big and how old the observable part of the universe is. We know that the temperature at the centre of the Sun is about 15 million degrees Celsius. We know that the Earth has a roughly spherical core of molten iron. We know that the Earth is roughly, though not exactly, spherical, and that (with suitable caveats about moving frames of reference) our planet goes round the Sun rather than being fixed in space while the Sun goes round *it*. We know that many features of an animal's form are determined, to a significant degree, by a long, complicated molecule that lives inside the nucleus of its cells. We know that bacteria and viruses cause most of the world's diseases. We know that everything is made from seventeen fundamental particles.

'Know' is one of those simple yet difficult words. How can we know, to take a typical example, what the temperature is at the Sun's centre? Has anyone been there to find out?

Well, hardly. If scientists are right about the temperature at the centre of the Sun, nobody who was suddenly transported there would survive for a nanosecond. In fact, they'd burn up long before they even reached the Sun. We haven't sent measuring instruments to the centre of the Sun, for the same reason. So how can we possibly know how hot it is at the centre, when no person or instrument can be sent there to find out?

We know such things because science is not limited to just *observing* the world. If it were, it would be firmly back in the human-centred realm. Its power derives from the possibility of thinking about the world, as well as experiencing it. The main tool of science is logical inference: *deducing* features of the world from a combination of observation, experiment and theory. Mathematics has long played a key role here, being the best tool we currently have for making quantitative inferences.

Most of us understand in broad terms what an observation is: you take a look at things, you measure some numbers. Theories are trickier. Confusingly, the word 'theory' has two distinct meanings. One is 'an idea about the world that has been proposed, but has not yet been tested sufficiently for us to have much confidence that it is valid'. A lot of science consists of proposing theories in this sense, and then testing them over and over again in as many ways as possible. The other meaning is 'an extensive, interconnected body of ideas that have survived countless independent attempts at disproof'. These are the theories that inform the scientific worldview. Anyone who tries to convince you that evolution is 'only a theory' is confusing the second use with the first, either through intention to mislead, or ignorance.

There is a fancy word for the first meaning: 'hypothesis'. Few people actually use this because it always sounds pedantic, although 'hypothetical' is familiar enough. The closest ordinary word to the second meaning is 'fact', but this has an air of finality that is at odds with how science works. In science, facts are always provisional. However, well-established facts – well-developed and well-supported theories – are not *very* provisional. It takes a lot of evidence to change them, and often a change is only a slight modification.

Occasionally, however, there may be a genuine revolution, such as relativity or quantum theory. Even then, the previous theories often survive in a suitable domain, where they remain accurate and effective. NASA mostly uses Newton's dynamics and his theory of gravity to compute the trajectories of spacecraft, not Einstein's. An exception is the GPS system of navigational satellites, which has to take relativistic dynamics into account to compute accurate positions.

Science is almost unique among human ways of thinking in not only permitting this kind of revisionism, but actively *encouraging* it. Science is consciously and deliberately universe-centred. That is what the 'scientific method' is about. It is like that because the pioneers of science understood the tricks that the human mind uses to convince itself that what it *wants* to be true *is* true – and took steps to combat them, rather than promoting them or exploiting them.

There is a common misconception of the scientific method, in which it is argued that there is no such thing because specific scientists stuck to their guns despite apparent contrary evidence. So science is just another belief system, right?

Not entirely. The mistake is to focus on the conservatism and arrogance of individuals, who often fail to conform to the scientific ideal. When they turn out to have been right all along, we hail them as maverick geniuses; when they don't, we forget their views and move on. And that's how the real scientific method works. All the other scientists keep the individuals in check.

The beauty of this set-up is that it would work even if *no* individual operated according to the ideal model of dispassionate science. Each scientist could have personal biases – indeed, it seems likely that they do – and the scientific process would still follow a universe-centred trajectory. When a scientist proposes a new theory, a new idea, other scientists seldom rush to congratulate him or her for such a wonderful thought. Instead, they try very hard to shoot it down. Usually, the scientist proposing the idea has already done the same thing. It's much better to catch the flaw yourself, before publication, than to risk public humiliation when someone else notices it.

In short, you can be objective about what everyone *else* is doing, even if you are subjective about your own work. So it is not the actions of particular individuals that produce something close to the textbook scientific method. It is the overall activity of the whole community of scientists, where the emphasis is on spotting mistakes and trying to find something better. It takes only one bright scientist to notice a mistaken assumption. A PhD student can prove a Nobel prize-winner wrong.

If at some future date new observations conflict with what we think we know today, scientists will – after considerable soul-searching, some stubborn conservatism, and a lot of heated argument – revise their theories to resolve the difficulties. This does not imply that they are merely making everything up as they go along: each successive refinement has to fit more and more observations. The absence of complete certainty may seem a weakness, but it is why science has been so successful. The truth of a statement about the universe does not depend on how strongly you believe it.

Sometimes an entire area of science can become trapped in a massive conceptual error. A classic instance is 'phlogiston'. The underlying scientific problem was to explain the changes that occur in materials when they burn. Wood, for instance, gives off smoke and flame, and turns into ash. This led to the theory that wood *emits* a substance, phlogiston, when it burns, and that fire is made from phlogiston.

Volume 2 of the first edition of *Encyclopaedia Britannica*, dated 1771, says: 'Inflammable bodies … really contain the element fire as a conſtituent principle … To this ſubſtance … chemiſts have aſſigned the peculiar title of the Phlogiſton, which is indeed no other than a Greek word for the inflammable matter … The inflammability of a body is an infallible ſign that it contains a phlogiſton …' The same edition considers 'element' to mean earth, air, fire or water, and it has a fascinating analysis of the size of Noah's Ark, based on its need to contain only a few hundred species.

As chemists investigated gases, and started weighing substances, they made a discovery that spelt doom for the phlogiston theory. Although ash is lighter than wood, the total weight of all combustion products – ash, gas and especially steam – is greater than that of the original wood. Burning wood *gains* weight. So, if it is emitting phlogiston, then phlogiston must have negative weight. Given enough imagination, this is not impossible, and it would be very useful as an antigravity device if it were true, but it's unlikely. The discovery of the gas oxygen was the clincher: materials burn only in the presence of oxygen, and when they do, they take up oxygen from their surroundings. Phlogiston was a mistaken concept of 'negative oxygen'. In fact, for a time oxygen was referred to as 'dephlogisticated air'.

Significant changes in scientific orthodoxy often occur when new kinds of evidence become available. One of the biggest changes to our understanding of stars came when nuclear reactions were discovered. Before that, it seemed that stars ought to burn up their store of matter very rapidly, and go out. Since they visibly didn't, this was a puzzle. An awful lot of argument about the Sun's remarkable ability to *stay alight* disappeared as soon as scientists realised it shone by nuclear reactions, not chemical ones.

This discovery also changed scientists' estimate of the age of the solar system. If the Sun is a very large bonfire, and is still alight, it must have been lit fairly recently. If it runs on nuclear reactions, it can be much older, and by studying those reactions, you can work out *how* much older. The same goes for the Earth. In 1862 the physicist William Thompson (later Lord Kelvin) calculated that on the 'bonfire' theory, the planet's internal heat would have disappeared within 20-400 million years. His approach ignored convection currents in the Earth's mantle, and when these were taken into account by John Perry in 1895 the age of the planet was revised to 2-3 billion years. Following the discovery of radioactivity, George Darwin and John Joly pointed out in 1903 that the Earth had its own internal source of heat, caused by radioactive decay. Understanding the physics of radioactive decay led to a very effective method for dating ancient rocks ... and so it went. In 1956 Clair Cameron Patterson used the physics of uranium decaying into lead, and observations of these elements in several meteorites, to deduce the currently accepted age of the Earth: 4.54 billion years. (The material in meteorites formed at the same time as the planets, but has not been subjected to the same complicated processes as the material of the Earth. Meteorites are a 'frozen' record of the early solar system.)

Independent verification has come from Earth's own rocks; in particular, tiny particles of rock called zircons. Chemically, these rocks are zirconium silicate, an extremely hard material that survives destructive geological processes such as erosion, and even metamorphism, where rocks are heated to extreme temperatures by volcanic intrusions. They can be dated using radioactive decay of uranium and thorium. The most ancient zircons yet observed – small crystals found in the Jack Hills of Western Australia – are 4.404 billion years old. Many different lines of evidence all converge on a similar figure for the age of our planet. This is why scientists are adamant that contrary to the claims of Young Earth creationists, a 10,000-year-old planet is completely inconsistent with the evidence and makes absolutely no sense. And they have come to this conclusion not through belief, or by seeking only confirmatory evidence and ignoring anything that conflicts, but by *trying to prove themselves wrong*.

No other system of human thought has the same kind of self-scrutiny. Some come close: philosophy, the law. Faith-based systems do change, usually very slowly, but few of them advocate self-doubt as a desirable instrument of change. In religion, doubt is often anathema: what counts is *how strongly you believe things*. This is rather evidently a human-based view: the world is what we sincerely and deeply believe it to be. Science is a universe-based view, and has shown many times that the world is *not* what we sincerely and deeply believe it to be.

One of Benford's examples illustrates this point: James Clerk Maxwell's discovery of electromagnetic waves travelling at the speed of light, implying that light itself is a wave. Human-centred thinking could not have made this discovery, indeed would have been sceptical that it was possible: 'The poets' and philosophers' inability to see a connection between sloshing currents in waves and luminous

36

sunset beauty revealed a gap in the human imagination, not in real-ity,' Benford wrote.

Similarly, the Higgs boson, by completing the standard model, tells us that there is far more to our universe than meets the eye. The standard model, and much of the research that led to it, starts from the idea that everything is made from atoms, which is already far removed from everyday experience, and takes it to a new level. *What are atoms made of?* Even to ask such a question, you have to be able to think outside the box of human-level concerns. To answer it, you have to develop that kind of thinking into a powerful way to find out how the universe behaves. And you don't get very far until you understand that this may be very different from how it appears to behave, *and* from how human beings might want it to behave.

That method is science, and it occupies the second of Benford's categories: the universe as a context for humanity. In fact, that is where its power originates. Science is done by people, for people, but it works very hard to circumvent natural human thought-patterns, which are centred on *us*. But the universe does not work the way we want it to; it does its own thing and we mostly go with the flow. Except that, being part of the universe, we have evolved to feel comfortable in our own little corner of it. We can interact with little pieces of it, and sometimes we can bend them to our will. But the universe does not exist in order for us to exist. Instead, we exist because the universe is that kind of universe.

Our social lives, on the other hand, operate almost exclusively in Benford's first category: humans as a context for the universe. We have spent millennia arranging this, re-engineering our world so that things happen *because we want them to*. Too cold? Build a fire. Danger-ous predators? Exterminate them. Hunting too difficult? Domesticate useful animals. Get wet in the rain? Build a house with a roof. Too dark? Switch on the light. Looking for the Higgs? Spend €7.5 billion.

As a result, most of the things we now encounter in our daily lives have been made by humans or extensively modified by humans. Even the landscape has been determined by human activity. Britain's hills have been shaped by extensive ancient earthworks, and most of its forests were cut down in the Iron Age so that farms could exploit the land. That wonderful scenery that you find at places like the stately home of Chatsworth – 'nature in all its glory', with a river flowing between sweeping hills, dotted with mature trees? Well, most of it was constructed by Capability Brown. Even the Amazon rainforest now seems to be the result of agricultural and architectural activity by ancient South American civilisations.

The differences between the two Benfordian worldviews are profound, but remain manageable as long as they don't overtly clash. Trouble arises when both worldviews are applied to the same things. Then, they may conflict with each other, and intellectual conflict can turn into political conflict. The uneasy relation between science and religion is a case in point. There are comforting ways to resolve the apparent conflict, and there are plenty of religious scientists, although few of them are Biblical literalists. But the default ways of thinking in science and religion are fundamentally different, and even deter-mined social relativists tend to feel uneasy when they try to claim there's no serious conflict. Benford's distinction explains why.

Most religious explanations of the world are human-centred. They endow the world with purpose, a human attribute; they place humans at the pinnacle of creation; they consider animals and plants to be resources placed on Earth for the benefit of humanity. In order to explain human intelligence and will, they introduce ideas like the soul or the spirit, even though no corresponding organs can be found in the human body, and from there it is a short step to the afterlife, whose existence is based entirely on faith, not evidence. So it should be no surprise that throughout history, science and religion have clashed. Moderates in both camps have always understood that these clashes are in a sense unnecessary. Looking back after enough time has passed, it is often hard to understand what all the fuss was about. But at the time, those two distinct worldviews simply could not accommodate each other.

The biggest battleground, in this context, is life. The astonishing world of living organisms: Life with a capital L. And, even more so, human consciousness. We are surrounded by life, we ourselves are conscious living beings … and we find it all terribly mysterious. Thirty thousand years ago some humans could carve quite realistic animals and people from bone or ivory, but no one, even today, knows how to breathe life into an inanimate object. Indeed, the idea that life is something you can 'breathe into' an inanimate object is not particularly sensible. Living creatures are not made by starting with a dead version and bringing it to life. Universe-centred thinkers understand this, but human-centred thinkers often see the body – especially the human body – as a dead thing that is animated by a separate, and immaterial, soul or spirit.

The proof of course is that we observe the reverse process on a regular basis. When someone dies, life seems to pass from their body, leaving a corpse. Where did the life *go*?

Agreed, science doesn't fully understand what gives us our personalities and consciousness, but it is pretty clear that personality derives from the structure and operation of a brain inside a body, interacting with the external world, especially other human beings. The person develops as the human develops. It's not a supernatural *thing*, inserted at conception or birth, with a separate existence of its own. It's a process carried out by ordinary matter in a living person, and when that person dies, their process *stops*. It doesn't depart into a new existence outside the ordinary universe.

In a human-centred view, souls make sense. In a universe-centred one, they look like a philosophical category error. In centuries of studying human beings, not a shred of convincing scientific evidence has ever been found for a soul. The same goes for all of the supernatural elements of all of the world's religions. Science and religion can coexist peacefully, and it's probably best that they do. But until religions discard the supernatural, these two very different world-views can never be fully reconciled. And when fundamentalists try to discredit science because it conflicts with their beliefs, they bring their beliefs into disrepute and provoke unnecessary conflict.

However, even though human-centred thinking can be abused, we cannot understand our place in the universe by using *only* universe-centred thinking. It's a human-centred question, and our relationship to the universe involves both points of view. Even though everything in the universe is made from seventeen fundamental particles, it's how those particles are combined, and how the resulting systems behave, that make us what we are.

THREE

SEEPAGE BETWEEN WORLDS

 The button having been pressed, the Archchancellor had noticed, *not* for the first time, that Lord Vetinari had a most useful talent, which was to be extremely volcanically angry without even slightly losing his composure. *Corpses* would have admired the coldness that he could insert into the most innocent conversation.

But now, in mid-reverie, Mustrum Ridcully heard a scream emanating from the High Energy Magic building. The scream was very closely followed by a number of wizards. They seemed to be fleeing, but he grabbed one and held on tight.

'Here! Has something gone horribly wrong?'

'I should say so, sir! There's a *woman*! And she's *angry*!'

This last wail was larded with the inference that only an Archchancellor could deal with a very angry woman. Fortuitously, Mustrum Ridcully was the very Archchancellor they needed, because for one thing he knew how to soothe, but he also knew when to twinkle and – more importantly perhaps – he also knew when *not* to twinkle. This looked like it could be a vital skill in the case of this particular lady, who was standing in the entrance to the HEM with her arms akimbo and a definite look of annoyance, a look which was tinted with a palpable sense that there had better be an explanation and, moreover, an extremely *good* one.

The Archchancellor took care as he walked towards her, and at exactly the proper moment took off his hat and bowed, not too theatrically, with just the right amount of olde-worlde charm. 'Do excuse me, madam, how may I be of service?' he said courteously. 'I thought I heard a scream?'

She glowered at him. 'Oh, I am sorry, but I punched one of your chaps. Couldn't help it. Found myself where I shouldn't be and thought: When in doubt strike first. I *am* a librarian, you know. And who are you, sir?'

'Madam, my name is Mustrum Ridcully and I am Archchancellor of this college.'

'And what you don't know isn't knowledge, by any chance? No!' The woman watched Ridcully's face and realised that he was as bewildered as her. 'Don't answer that! Just tell me where I am and why. I can't get any coherence with all these men scuttling about like drones around a hive.'

'Madam, I quite understand your feelings myself – it takes ages to get any real coherence out of them. Alas, that is the curse of academia; but with the aforesaid in mind I *will* tell you that you appear to have magically landed in Unseen University, and have been caught up in what I might now call a "science" experiment, although it may seem to you to be like magic, and very hard to explain at the moment. I do have my suspicions as to how you got here, believe me, for my thaumometer nearly went off the scale a few minutes ago, suggesting that what we might call magic is wild.' He paused, before adding reassuringly, 'But don't fret; I'm used to this sort of thing. Running a university does have its ups and downs, and I truly believe I know the cause of this, and we will get it right as soon as possible. May I say that I would be very pleased if you would be our guest until that happy time.'

She looked at him askance, in a slightly dazed fashion, and said, 'Somehow I appear to have turned up mysteriously in something like Balliol College; it certainly reminds me of it, oh my word, yes.

Oh dear, where are my manners?' The woman held out her hand to Ridcully and said, 'How do you do, sir? My name is Marjorie Daw, you know, rather like the nursery rhyme?* And please, I don't know how I got here, I don't know how I can get back to where I belong … and I am feeling rather sick.'

While she was speaking, a white-robed wizard rushed to the side of Mustrum Ridcully and handed him a small piece of paper; then scuttled away quickly.

Mustrum glanced at what was on the paper. 'I believe, madam, that you hail from England, on planet Earth as you call it – a fact which I have just established, since my Librarian can't find any other place in the multiverse where that particular nursery rhyme should be sung.'

She stared at him, the words 'planet' and 'multiverse' rocketing into her brain, back out again and then – because she *was* a librarian – pulling out an index card or two and settling in again for a nice bit of research. Then she crumpled gently downwards towards the lawn, where she was gallantly caught by the Archchancellor.

She came round in a matter-of-fact way, saying, 'Sorry, there must be something about this travelling that doesn't suit me.' Her eyes narrowed and her lip curled as she continued, 'It won't happen again, I *assure* you.'

Ridcully, apparently lost in admiration for this surprisingly amazing woman, led her to the office of Mrs Whitlow, the housekeeper, who very shortly afterwards reported back that the mysterious lady was snoring in the best guest room available. And Mrs Whitlow also gave the Archchancellor a *look* – one of those looks that spoke for themselves – for after all, he had just carried a lady into the university. It concluded that, well, presumably a man could do anything he wanted in his *own* university, but please let there be no hanky or panky or, even worse, *spanky.*

---

* She had rather liked the name until she went to school; the other kids teased her until one day she took umbrage and there was an up-and-downer, after which they showed some respect.

Mustrum Ridcully, on the other hand, did not go to sleep immediately, but instead, once all the guests and visitors had gone, ambled along to the university library, where he spoke to the head Librarian, who promptly carried out the task that Ridcully had presented to him.

Although he had a very pointy hat and, on special occasions, wore very ornate robes, Mustrum Ridcully was also very smart. Smart was a necessary part of life in the university if you wanted to *have* a life in the university. He prided himself on his memory for small things, so within the hour he headed to the study of Ponder Stibbons. He was followed dutifully by the Librarian, whose skill at picking up data fast was legendary.

'Simian* and gentlemen,' Ridcully summarised, 'I am convinced that the Great Big Thing so recently put to work by the wizards of the High Energy Magic building may have struck what I am reliably informed is called a *hitch* ... yes, Mister Stibbons?'

Everyone knows that if you have foolishly done something wrong then your first step must be to determine if the blame can be laid elsewhere, but Archchancellor Ridcully knew where *all* the bucks stopped, and so Ponder's best defence would therefore be to state a clear intention to return the world, as soon as possible, to the *status quo ante*, and by any means necessary.

'On a point of order, Archchancellor,' the wizard replied, 'the word in question is a *glitch* and, as they go, a not particularly bad one, given that as far as we can tell nobody has been injured, I'm pleased to say. According to HEX, Archchancellor, your surmise that we are cross-linked to Roundworld is correct. Well done, sir! Finding a clue in that children's rhyme was an amazing surmise. Unfortunately, it also makes me worry that there may be more ... *seepage* between our worlds ...'

---

* The Librarian of Unseen University, who gets a capital 'L', is an orangutan, because of an accident when a spell escaped from a book of magic. See *The Light Fantastic*.

Ridcully frowned. 'Mister Stibbons. We have meddled in Round-world rather too many times, in my opinion. In fact, as I recall, it was the Dean who caused the place to come into being, don't you remember? He was mucking about with some firmament, so techni-cally speaking he created the place. Mind you,' he went on, 'I think it would be a very good idea if we don't let anybody know about *that*. There would be no end of arguments.'

Ponder nodded vigorously.

Ridcully grinned and continued thoughtfully, with a certain amount of malice aforethought, 'It does seem to me, Mister Stibbons, that we should send an agent in there to see how things are. After all, Miss Daw has stumbled into our world, and therefore we have a duty to see nothing untoward is happening in hers as a result of your … experiment. Indeed, in the interests of all concerned, I think we should definitely send someone the other way. We *are* responsi-ble for the place.' Mustrum Ridcully stroked his beard, a signal to all who knew him that he was feeling rather nasty and mysterious. 'I think, yes, the Dean himself ought to go and have a look around.' The beard was stroked again, and Ridcully continued, 'For backup, you had better send Rincewind with him; he's been looking a little peaky lately, so a change of air will do him good.'

'Alas, sir,' said Ponder, 'if you recall, and I *know* you recall, the Dean is now Archchancellor of Pseudopolis University, and we haven't inducted a new Dean yet.'

Undeterred, Ridcully said, 'Get him anyway! He was the one who created Roundworld. He can't just shrug it off; he ought to see how the old place is doing. Send him a clacks. We need action today. We want no more seepages!'

FOUR

WORLD TURTLES

Before the Large Hadron Collider was turned on, there were attempts to get court injunctions to stop it, in case it created a mini black hole and gobbled up the universe. This was not *totally* silly, but it ignored a worse problem: according to the cosmological theory of eternal inflation, any part of the universe could blow up at any moment – see chapter 18.

Thanks to the switching on of the Great Big Thing, Marjorie Daw has seeped into Discworld. Since she is a librarian, we suspect the seepage happened through L-space, the interconnected space of all libraries that ever have existed or ever could exist.

This may not be the first time something has seeped from Roundworld into Discworld either. Long ago, when the Omnian religion was founded, its adherents came to believe that Discworld, belying its name, is actually round. Where did *that* idea come from? For that matter, how did many early Roundworld cultures get the complementary idea that their world is flat?

We can gain some knowledge about early human beliefs from archaeology, the branch of science that examines evidence from our past. The artefacts and records that survive give us clues about how the ancients thought. Those clues can to some extent be clarified by another branch of science, psychology: the study of how people think. The pictures that emerge from the combination of these two sciences are necessarily tentative, because the evidence is indirect. Scholars can, and do, have a field day arguing about the interpretation of a cave painting or a stick with marks on it.

Ancient myths and legends possess a number of common features. They often focus on deep, mysterious questions. And they generally answer those questions from a human-centred viewpoint. The Discworld series takes Roundworld mythology seriously, to humorous effect; nowhere more so than in its basic geography and its magical supports – elephants and turtle. Here we'll take a look at how various ancient cultures imagined the form, and purpose, of our world, looking for common elements and significant differences. Especially flat worlds and world-bearing animals. Here elephants turn out to be particularly problematic, most likely a case of mistaken identity. In chapter 20 we revisit some of these ancient myths, which will illuminate the science of human belief systems.

In a human-centred view, a flat world makes more sense than a round one. Superficially the world looks flat, ignoring mountains and suchlike and concentrating on the big picture. In the absence of a theory of gravity, people assumed that objects fell *down* because that was their natural resting-place. To prove it, just lift a rock off the ground and let go. So a round world seems implausible: things would fall off the bottom half. In contrast, there's no danger of falling off a flat world unless you get too close to the edge.

There is one effective way to counteract this natural tendency to fall downwards: place something underneath as a support. This support may in turn need something underneath to support *it*, but you can iterate the process many times provided ultimately everything rests on something firm. This process, known as building, was effective enough to erect the Great Pyramid of Khufu at Giza, built in 2560 BC and over 145 metres high. It was the tallest building in the world until 1300, when the architect of Lincoln cathedral cheated by using a lot more up and a lot less sideways.

A common feature of human-centred thinking is that it often works well until you start to ask questions that transcend the human scale. Then it has a habit of falling to pieces. The line of thought just described seems fairly foolproof until you go for the big picture. Applying the kind of logical reasoning that drives so many Discworld stories, it is impossible not to ask: *What keeps the world up?* Human-centred thinking provides an obvious and compelling answer: something supports it. In Greek mythology, it was Atlas, bearing the world on his sturdy shoulders. Discworld sensibly plumps for a more plausible support cast: the giant world-bearing elephant. As belt and braces, there is not just one of them, but four – or possibly five, if the legend recounted in *The Fifth Elephant* is to be believed.

All well and good, but both universe-centred science and human-centred myth-making can hardly fail to ask a supplementary question: *What keeps the elephants up?* If the idea of an ordinary elephant hovering in mid-air is ludicrous, how much more so is that of a vast, extraordinarily heavy elephant doing the same? Discworld's answer is A'Tuin, a giant space-faring turtle. The turtle's shell provides a firm place for the elephants to stand. As a cosmology, it all hangs together pretty well ... but of course a further question arises: *What keeps the turtle up?*

It might seem that we could go on like this indefinitely, but at this point observations of nature come into play. The natural world provides a long list of exceptions to the belief that the natural place of any object is on the ground: celestial bodies, clouds, birds, insects and all water-borne creatures – fish, crocodiles, hippos, whales and, crucially, turtles.

However, we can prune the list. Birds and insects do not remain aloft indefinitely; wait long enough and they do, in fact, descend to their natural place, typically a tree or a bush. The Sun, Moon and stars do not inhabit the terrestrial realm at all, so there is no reason to expect them to behave in a human-centred way – and they don't. Assigning them to the realm of the supernatural has so many attractions that it becomes virtually unavoidable. The same arguably goes for clouds, which have a habit of producing awe-inspiring phenomena such as thunder and lightning. Scratch clouds. Crocodiles and hippos are out: they spend a lot of time on land. Fish are not renowned for doing that, but no sensible person would try to fit four elephants on top of a fish.

Which leaves turtles.

Small turtles spend a lot of time on rocks, but no one in their right mind would expect a small turtle to hold up four giant world-bearing elephants. Big turtles come out onto land to lay their eggs, but that's a mystical event and it doesn't cast serious doubt on the theory that a turtle's natural place is in water. Where, please notice, *it does not require support.* It can swim. So it stands to reason that any self-respecting giant space-faring turtle will swim through space, which implies that it needs no artificial support to avoid falling. Examining the animal more closely, a world-spanning turtle seems ideal as a support for giant elephants. It is hard to imagine what could perform the task better.

In short, Discworld is, as stated earlier, the sensible way to make a world.

By comparison, Roundworld makes no sense. It's the wrong shape, it's held up by nothing, and it swims through space unaided despite not being the right shape to swim through *anything.* Basically, it's a giant rock, and you all know what rocks do when you throw them in the lake. It is hardly surprising that it took the wizards a long time to come to terms with the way Roundworld organises itself. Accordingly, we should not be surprised to find that pre-scientific humanity had the same problem.

Flat worlds, giant elephants, world-bearing turtles ... how did these enter the human psyche? One of the ironies of human-centred thinking is that it is unavoidably attracted to superhuman questions – the big picture. What are we? Why are we here? Where did it all come from? And one of the ironies of universe-centred thinking is that it is far better equipped to answer human-scale questions than cosmic ones.

If you want to find out how the rainbow gets its colours, you can pass light through a glass prism in a darkened room. This is what Isaac Newton did in about 1670, though he had to overcome some practical problems. The worst was his cat, which kept wandering into the attic to find out what Isaac was doing, pushing open the door and letting light in. So the ingenious scientist cut a hole in the door and nailed up a piece of felt, inventing the cat flap. When puss had kittens, he added a smaller hole next to the big one, which probably seemed logical at the time.* Anyway, once the feline disruption was taken care of, Newton discovered that white light from the Sun splits into colours, and optics was born.

This kind of experiment is a cinch for things like light, which can be confined to a laboratory (if the cat complies). If you want to discover the nature of the universe, however, it's not so easy. You can't put the universe on a laboratory bench, and you can't step outside it to observe its form, or go back in time to see how it began. The wizards can do, and have already done, all of these things; however, neither the scientists nor the theologians of Roundworld are likely to accept that the Dean of Unseen University kicked it all off by poking his finger in.

Instead, human-centred thinkers on Roundworld tend to go for human-level explanations like emperors and elephants, scaled up to superhuman levels to become gods and world-bearers. Most human civilisations have a creation myth – often several, not always compatible. Universe-centred thinkers have to fall back on scientific inference, and test the resulting theories indirectly. Their cosmological scenarios have often fared little better than most creation myths. Some look remarkably similar: compare the Big Bang to Genesis.

---

* Like all really nice stories this tale, told by a 'country parson', may be false. Other versions say that Newton kept losing time from his research by letting the cat out. Selig Brodetsky's *Sir Isaac Newton* and Louis Trenchard More's *Isaac Newton: a Biography* both state that the great mathematician did not allow either a cat or a dog to enter his chamber. But in 1827 J.M.F. Wright, who lived in Newton's former rooms at Trinity College, Cambridge, wrote that the door once had two holes – by then filled in – of the right size for a cat and a kitten.

However, scientific cosmologists do try to prove themselves wrong, and keep looking for weaknesses in their theories even when observations seem to confirm that they're right. Typically, after about twenty years of increasingly good supporting evidence, these theories start to unravel as the observations become more sophisticated: see chapter 18.

Our ancestors needed to rationalise the things they observed in the natural world, and creation myths played a significant role. It can therefore be argued that they helped to bring about today's science and technology, because they long ago drew humanity's attention to the big questions, and held out hope of answering them. So it's worth examining the similarities and differences between the creation stories of different cultures – especially when it comes to world-bearing elephants and space-faring turtles. Along with a third common world-bearing creature, the giant snake.

The world turtle (cosmic turtle, divine turtle, world-bearing turtle) can be found in the myths of the Chinese, Hindus and various tribes of native North Americans, in particular the Lenape (or Delaware Indians) and the Iroquois.

Around 1680 Jasper Danckaerts, a member of a Protestant sect known as Labadists, travelled to America to found a community, and he recorded a Lenape myth of a world turtle in *Journal Of A Voyage To New York In 1679-80*. We paraphrase the story from a 1974 article by Jay Miller.* At first, all was water. Then the Great Turtle emerged, mud on its back became the Earth, and a great tree grew. As it rose skywards, one twig became a man; then it bent to touch the Earth and another twig became a woman. All humans descended from these two. Miller adds: 'my ... conversations with the Delaware indicate that life and the Earth would have been impossible without the turtle supporting the world.'

---

* Jay Miller, Why the world is on the back of a turtle, *Man* **9** (1974) 306–308.

According to the Iroquois creation story, immortal Sky People lived on a floating island before the Earth existed. When one of the women discovered that she was going to have twins, her husband lost his temper and pulled up a tree at the island's centre, the tree being their source of light at a time when the Sun did not exist. The woman looked into the hole thus created, and far below she saw the ocean that covered the Earth. Her husband pushed her into the hole, and she fell. Two birds caught her, and tried to get mud from the ocean floor to make land for her to live on. Finally Little Toad brought up mud, which was spread on the back of Big Turtle. The mud grew until it turned into North America. Then the woman gave birth. One son, Sapling, was kind, and filled the world with all good things; the other, Flint, ruined much of his brother's work and created everything evil. The two fought, and eventually Flint was banished to live as a volcano on Big Turtle's back. His anger can still sometimes be felt when the Earth shakes.

In these myths, there are partial parallels with ancient Egyptian mythology, in which the primal mound or *benben* rose from a sea of chaos. The god Seth wanted to kill his brother Osiris. He constructed a coffin, lured Osiris inside, shut the lid, sealed it with lead and threw it in the Nile. Their sister Isis set out to find Osiris, but Seth got there first and cut him into 14 pieces. Isis located 13 of them, but a fish had eaten Osiris's penis. So she made an artificial one for him from gold, and sang until he came back to life.

The world-bearing turtle never made it into the Egyptian pantheon, but it was common in ancient central America, among cultures such as the Olmecs. To many of these cultures, the world was both square and round, and it was also a caiman or turtle floating on a primordial sea, which represented the Earth and might or might not carry it. The world had four corners, one for each cardinal direction, and a fifth symbolic point at its centre. The cosmos was divided into three horizontal layers: the underworld below, the heavens above, and the everyday world in between.

In another central American culture, the Maya civilisation, thirteen creator gods constructed humanity from maize dough. The world was carried at its four cardinal points by four *bacabs*, elderly deities of the earth's interior and waters, shown carrying a sky-dragon in early depictions but later believed to be drowned ancestors. Their names were Cantzicnal, Hobnil, Hosanek and Saccimi, and each ruled one of the four directions.* They were closely associated with four rain gods and four wind gods. They can appear as a conch, a snail, a spider web, a bee-like suit of armour, or a turtle. In the *Dresden Codex* the turtle is also associated with the rain-god Chaac, which similarly has four aspects, one for each cardinal direction.

At the Puuc Maya site at Uxmal there is a building called the House of the Turtles, whose cornice is decorated with hundreds of the animals. Its function is unknown, but the Maya associated turtles with water and earth. Their shells were used in making drums, and seem to have been associated with thunder. The god Pauahutun, who like Atlas carried the world on his shoulders, is sometimes shown wearing a turtle-shell hat. The Maize God is occasionally shown emerging from a turtle's shell. The Mayan name for the constellation Orion is Ak'Ek' or Turtle Star.

The *Popol Vuh* of the Quiché Maya provides more detail. It tells of three generations of deities, beginning with the creator grandparents of the sea and the lightning gods of the sky. The Mayans were corn farmers, so their human-centred worldview naturally related to the cycle of wet and dry seasons: their creator gods brought rain and the corn cycle into being. Their gods came as a standard package. Each god was associated with an aspect of the Mayan calendar, so one function of the calendar was to specify which god was in the ascendant at any given time. Often gods possessed several different aspects, and some of the major deities had four aspects, one for each cardinal direction, each with slightly different responsibilities.

* It's fascinating how priests always know the *names* of the gods.

The *Popol Vuh* tells that before the Earth appeared, the universe was a huge freshwater sea, above which was a blank sky with no stars or Sun. In the sea dwelt the creator grandparents Xpiyacoc and Xmucane. Below was Xibalba, the place of fright, domain of the gods One Death and Seven Death. The gods of sea and sky decided that they would create people to worship them. Since such creatures would need somewhere to live, the gods created the Earth, raising it from the primordial sea and covering it in vegetation.

That was Mayan cosmogony: the origin of the universe. In their cosmology (the shape and structure of the universe) the Earth was a flat disc, but it also had aspects of a square, whose corners were determined by the rising and setting of the solstice Sun, and whose sides were four great mythical mountains. It has been suggested that the notion of a square world reflects the shape of a cornfield. A rope formed a protective perimeter, reminiscent of Discworld's 'circumfence'* but this one was intended to keep out malevolent supernatural beings. Each of the mountains was the home of one aspect of a grandfather deity, Mayan name unknown or uncertain, referred to by anthropologists as God N. The gods' homes could be reached through caves, but these created gaps in the protective perimeter so that evil could enter.

The Earth was next made ready for growing corn. So the children and grandchildren of the creator grandparents, now living on the Earth, set up the Sun and the yearly cycle of the seasons and synchronised them with the movements of the Moon and Venus. There were two children, Hun Hunahpu and Vucub Hunahpu. The first married Bone Woman – the book does not say how she came into being (just as Genesis tells us that Cain's wife 'dwelt in the land of Nod', but is silent about the creation of both Nod and the wife). When Bone Woman died, Hun Hunahpu and Vucub Hunahpu went to the underworld, suffering defeat at the hands of the two lords of

---

* A 10,000-mile long drift net built to catch items falling over the edge.

death. Blood Woman, the daughter of an underworld being, was made pregnant by spittle from Hun Hunahpu's dead head, and gave birth to Hunahpu and Xbalanque, the hero twins. Much of the tale deals with the twins' eventual defeat of the lords of death, which required assistance from their grandparents. Xmucane made a mixture of corn and ground bone into dough, from which the creator grandparents formed the first people. Job done, the hero twins became the Sun and the full Moon.

God N is often shown wearing a net bag on his head. One of his manifestations was as a possum; another was as a turtle. An inscribed stone at Copán bears his name 'yellow turtle', in the form of his image together with phonetic signs for *ak* – meaning turtle. In his turtle aspect, God N represented the Earth, because the creation of the Earth, rising from the primordial sea, was like a turtle coming to the surface of a pool. God N also manifested himself as the four *bacabs*, whom the sixteenth-century Bishop of Yucatán Diego de Landa described as 'four brothers whom [the creator] god placed, when he created the world, at the four points of it, holding up the sky so that it should not fall'.

Benford's distinction is very visible here. The Mayan view, like that of many ancient cultures, was human-centred. They tried to understand the universe in terms of their own everyday experiences. Their stories rationalised nature, by portraying it in human terms – only bigger. But within that framework, they did their best to tackle the big questions of life, the universe, and everything.

To westerners, a turtle/elephant world is most commonly associated with Hinduism. Turtles are often confused with tortoises, as they generally are in American English. Philosopher John Locke's *Essay Concerning Human Understanding* in 1690 mentions an 'Indian who said the world was on an elephant which was on a tortoise'. In his 1927 *Why I Am Not A Christian* Bertrand Russell writes of 'the Hindu's view, that the world rested upon an elephant and the elephant rested upon a tortoise', adding, 'When they said, "How about the tortoise?" the Indian said, "Suppose we change the subject."' The elephant-turtle story remains in common circulation, but it is a misrepresentation of Hindu beliefs, conflating two separate mythical beings: the world-turtle and the world-elephant. In fact, Hindu mythology features three distinct species of world-bearing creature: tortoise, elephant and snake, with the snake being arguably the most important.

These creatures occur in several guises. The commonest name for the world-tortoise is Kurma or Kumaraja. According to the *Shatpatha Brahmana* its upper shell is the heavens, its lower shell the Earth, and its body is the atmosphere. The *Bhagavata Purana* calls it Akupara – unbounded. In 1838 Leveson Vernon-Harcourt published *The Doctrine of the Deluge*, whose purpose is clearly indicated by its subtitle: *vindicating the scriptural account from the doubts which have recently been cast upon it by geological speculations.* In it, he wrote of a tortoise called Chukwa that supported Mount Meru. This mountain is sacred in both Hindu and Buddhist cosmology, the centre of the universe – physical, spiritual and metaphysical – where Brahma and the demigods reside. Vernon-Harcourt attributes the story to an astronomer who described it to Bishop Heber 'in the Vidayala school in Benares'. Since the word 'vidyayala' (note slight difference in spelling) means 'school' in Sanskrit, it is hard to give the report much credit. *Brewer's Dictionary of Phrase and Fable* includes the entry '*Chukwa*. The tortoise at the South Pole on which the Earth is said to rest', but there is little evidence to support this statement. However, Chukwa appears in the Ramayana as the name of a world-elephant, also known as Maha-padma or -pudma. Most likely various mythological entities were being confused and their stories combined.

Some sources say that Chukwa is the first and oldest turtle, who swims in the primordial ocean of milk and supports the Earth. Some also say that the elephant Maha-Pudma is interposed. This story apparently occurs in the *Puranas*, dating from the Gupta period (320-500). Whether the Hindus *believed* this myth, other than in a ritual sense, is debatable. Hindu astronomers of the Gupta period knew the Earth was round, and they may even have known that the Earth goes round the Sun. Perhaps there were 'priests' and 'scientists' – human- and universe-centred thinkers – then too.

The ocean of milk appears in one of the most famous reliefs at one of the great world heritage sites, the Khmer temple complex of Angkor Wat in Cambodia. In one version of Hindu cosmology, the ocean of milk was one of seven seas, surrounding seven worlds in concentric rings. Horace Hayman Wilson's 1840 translation of the *Vishnu Purana* relates that the creator god Hari (*aka* Vishnu and Krishna) instructed all the other gods to throw medicinal herbs into the sea of milk, and to churn the ocean to make *amrit* – the food of the gods. Assorted gods were told to use the mountain Mandara as a churning-stick, winding the serpent Vásuki round it like a rope. Hari himself, in the form of a tortoise, served as a pivot for the mountain as it was whirled around.

Around 1870 Ralph Griffith translated the *Rámáyan of Válmíki* into verse. Canto 45 of Book 1 relates that it didn't go as well as had been hoped. When the gods and demons continued to churn the Ocean of Milk, a fundamental engineering blunder became apparent:

> *Mandar's mountain, whirling round.*
> *Pierced to the depths below the ground.*

They implored Vishnu to help them 'bear up Mandar's threatening weight'. Obligingly, he came up with the perfect solution:

> *Then Vishnu, as their need was sore,*
> *The semblance of a tortoise wore,*
> *And in the bed of Ocean lay*
> *The mountain on his back to stay.*

Despite its neglect in Discworld cosmology, we must now introduce another species of world-bearing animal: the snake.

You'll see why in a moment.

In many Hindu and Buddhist temples, the handrails of stair-cases are long stone snakes, which terminate at the lower end as a many-headed king cobra, each head having an extended hood. This creature is called a *naga*. The *nagas* of Angkor generally have seven heads in a symmetric arrangement: one in the centre, three either side. A Cambodian legend tells of the *naga* as a race of supernatural reptiles whose kingdom was somewhere in the Pacific Ocean; their seven heads correspond to seven distinct races, mythically associated with the seven colours of the rainbow.

The *Mahabharata* takes a fairly negative view of *nagas*, portraying them as treacherous and venomous creatures, the rightful prey of the eagle-king Garuda. But according to the *Puranas* the king of the *nagas*, Shesha (*aka* Sheshanag, Devanagari, Adishesha), was a creator deity. Brahma first saw him in the form of a devoted human ascetic, and was so impressed that he gave him the task of carrying the world on his head. Only then did Shesha take on the aspect of a snake, slith-ering down a hole in the Earth to reach the base of the world, so that instead of placing the planet on his head, he placed his head beneath the planet. As you would.

Why are we talking about world-bearing snakes, not exactly prom-inent in the Discworld canon?

World-bearing elephants are probably snakes that got lost in translation.

The Sanskrit word *naga* has several other meanings. One is 'king cobra'. Another is 'elephant' – probably a reference to the animal's snake-like trunk. Although world-bearing elephants appear in later Sanskrit literature, they are conspicuously absent from the early epics. Wilhelm von Humboldt has suggested that the myths of world elephants may have arisen from confusion between different mean-ings of 'naga', so that stories about the world-bearing serpent became corrupted into myths about world-bearing elephants. This is, in any case, an attractive idea for a culture that routinely used elephants for heavy lifting.

Classical Sanskrit writings include many references to the role of world elephants in Hindu cosmology. They guard and support the Earth at its four cardinal points, and the Earth shakes when they adjust their positions – an imaginative explanation for earthquakes. They variously occur as a set of four, eight, or sixteen. The *Amarakosha*, a dictionary in verse written by the scholar Amarasinha around AD 380, states that eight male and eight female elephants hold up the world. It names the males as Airavata, Anjana, Kumunda, Pundarika, Pushpa-danta, Sarva-bhauma, Supratika and Vamana. It is silent about the names of the females. The *Ramayana* lists just four male world elephants: Bhadra, Mahápadma, Saumanas and Virúpáksha.

It may or may not be significant that the name Mahápadma is mentioned in *Harivamsa* and *Vishnu Purana* as a supernatural snake. Like dragons in the mythology of other cultures, it guards a hoard of treasure. *Brewer's Dictionary* describes a 'popular rendition of a Hindu myth in which the tortoise Chukwa supports the elephant Maha-pudma, which in turn supports the world'. This variant spelling seems to come from a misprint in a 1921 edition of one of the stories of the *Mahabharata* by the Indian freedom fighter and poet Sri Aurobindo:

> *On the wondrous dais rose a throne,*
> *And he its pedestal whose lotus hood*
> *With ominous beauty crowns his horrible*
> *Sleek folds, great Mahapudma; high displayed*
> *He bears the throne of Death.*

However, this creature is clearly a giant cobra – unless you think the lotus hood is the elephant's ears.

The acorn and the oak have a superficially simple story, which we all understand: plant the acorn, water it, give it light, and it grows into the oak. However, that simple story cloaks a really difficult explanation of an immensely complex development: it is, in fact, much the same account as getting you from an egg. And there's another complication: not only does the oak come from the acorn: the acorn's origin is the oak. This is exactly like the chicken and egg cliché. The important question, though, is not 'which came first?'. That's a silly question, because they are both part of the repeating system. It's clear that the chicken is only the egg's way of making another egg. Before chickens, the same egg lineage used jungle fowl instead to make more eggs; long before that, it used little dinosaurs to make its eggs; and long before that, it used ancient amphibians.

The big problem with 'turtles all the way down' as an explanation is not the ludicrous mental image, amusing though that may be. Each turtle is indeed supported by the one beneath. The problem is how and why an entire infinite pile of turtles should exist. What matters in recursive systems is not which part came first, but the origin of *the whole system.* For eggs and their chickens, that story is mainly an evolutionary one, a sequence of developments that change progressively, so that now we have chickens when previously we had jungle fowl or dinosaurs. In this case, the origins of the system go all the way back to the first eggs, the first multicellular creatures that used embryonic development from eggs as part of their reproductive process. In that same way, the acorn is the modern version of a seed that used to produce early seed plants, and prior to that produced tree-ferns … all the way back to the origins of multicellular plants.

What we mean by 'immensely complicated development' also takes a bit of explaining. It's clear that the acorn doesn't *become* the oak tree, any more than the egg that generated you became you. The oak tree is mostly made from carbon dioxide extracted from the air, water from the soil and minerals, including nitrogen, also from the soil. In trees, those ingredients mainly make carbohydrates, cellulose and lignin, along with proteins for the working chemical machinery. The amount of material contributed by the acorn is minuscule. Similarly, almost all of the baby that (in a very restricted sense) became you, was built from a variety of chemicals obtained from your mother through the placenta. The tiny egg contributed very little by way of materials … but an awful lot by way of organisation. The egg functioned to recruit the chemicals that your mother provided, initiating and controlling the succession of stages – blastocyst, embryo, fetus – that led to your birth. Similarly, the acorn is already an embryo, and it has a very complex organisation, beautifully crafted to drive a root down into the soil, to extend leaves up into the air, and to start the business of becoming a tiny oak.

It's that word 'becoming' that we all have trouble with. Jack, on a hospital ethics committee, once had to explain how an embryo → fetus → baby → *becomes* human. It's not like switching on a light, he explained; it's more like painting a picture, or writing a novel. There isn't one paintbrush-stroke or one word that completes the task; it's a gradual becoming. 'That's fine,' a lay member of the committee replied, 'but how far into a pregnancy is it before you have a human being, not just an egg?' We seem to need to draw lines, even when nature fails to present us with tidily distinguished stages.

So let's not start with complex development, like acorns and eggs, when thinking about origins. Let's start with something genuinely simpler: a thunderstorm. Before the storm, there is a time of cool, clear skies, clouds moving with the wind, probably a weather front. What we don't see, because it's invisible, is the static electricity building up in the clouds. Clouds are masses of water droplets, billions of tiny spheres of liquid water in a mass of water vapour: a saturated solution of water in air. The droplets and vapour rise to the top of the cloud; then they fall back through the cloud, not quite dropping out as rain, and the cycle repeats. Many do drop out as rain when the storm starts, of course.

Clouds are very active structures, with massive circulations. They look gauzy and simple, but internally they are a mass of water-droplet and ice-particle currents. Each droplet and particle carries a tiny electric charge, and the cloud as a whole also acquires an electric charge, for much the same reason that your nylon underwear acquires an electrical charge opposite to that of your body. So the cloud has the opposite charge to that of the hills it passes over, a clear recipe for trouble. As the charge builds, the electric potential between the cloud and the ground gets bigger. Eventually it becomes big enough for lightning to make its own path between cloud and ground, following a trail of lower-resistance ionised air. Metal spikes sticking up from the ground, or on the top of tall buildings like churches, provide particularly good targets. In the absence of those, a person walking on a hill might be the unlucky Earth-end of a strike.

A thunderstorm seems simpler than an acorn becoming an oak, because it doesn't need lots of intricate organisation. But even a thunderstorm is not as simple as we tend to imagine: we don't know how the electric potential builds. There are 16 million thunderstorms each year on Roundworld, but we're still not really sure how they happen. No wonder we have trouble understanding how an acorn becomes an oak.

As for the *origin*, the beginnings of a storm, the beginnings of anything ... To explain thunderstorms, do we have to explain clouds? The constituents of the atmosphere? Static electricity? The elements of physics and physical chemistry? The origin of anything lies in the interactions of multiple causes. In practice, in order to explain the origin of a storm, or anything else, both the person providing the explanation and the one on the receiving end must have a lot of knowledge in common, covering many different areas. Unfortunately, it may not be present.

You might be an English teacher, an accountant, a housewife, a psychologist, a merchant, a builder, a banker or a student. The chances are that you will not have come across one or more phrases such as 'saturated solution' or 'particle carries a tiny electric charge'. And those phrases are themselves simplifications of concepts with many more associations, and more intellectual depth, than anyone can be expected to generate for themselves.

You might be a biology teacher, a mathematician, or even a science journalist, with a more extensive mental database in such areas. Even so we'd still have difficulty explaining the origin of storms, because *we* don't understand it in enough depth. None of us is a meteorologist. And even if we were, we *still* wouldn't be able to generate enough depth of understanding for you to be able to say, 'Ah, yes – I understand that now.' Jack is an embryologist, and understands eggs and acorns in some depth; he would have the same problem for the same reason, even for those examples. The origin of absolutely anything on Roundworld – of it, off it, all the way up to everything that exists – is a complicated mesh involving enormously many factors that we know very little about.

One way to duck out of this issue is to appeal to divine creation. If you believe in a creator god, you can invoke supernatural intervention to explain the origins of anything, from the universe to thunderstorms. Thor does a great job with his hammer: job done, thunder explained. Or don't you think so? We don't find that a very satisfactory explanation, because you then have to explain how the gods came to be, and where their powers came from. Maybe it's not Thor at all, but Jupiter. Maybe it's a giant invisible snake thrashing its coils. Maybe it's an alien spacecraft breaking the sound barrier.

Some quite sophisticated creation stories exist, as mentioned in chapter 4, but none of them are genuine explanations. The same form of words 'explains' absolutely anything, and would equally well appear to explain a lot of things that don't happen at all. If you think the sky is blue because God made it that way, you would be equally happy if it were pink, or yellow with purple stripes, and would offer exactly the same explanation. On the other hand, if you explain the colour of the sky in terms of light being scattered by dust in the upper atmosphere, and discover that the intensity of the scattered light is inversely proportional to the fourth power of the wavelength, then you will understand why short-wavelength blue light will dominate compared to the longer wavelengths of yellow and red. (The fourth power of a small number is very small indeed, and *inverse* proportionality means that small numbers are more important than big ones – just as one tenth is bigger than one hundredth.) Now you've learned something useful and informative, which you can apply to other questions.

However, this type of explanation only goes so far: it doesn't explain where the dust came from, or more difficult things like why blue light looks blue. If you want complete explanations of anything at all, creation is the way to go. Theology really does have all the answers. Indeed, the myriad religions and creeds on the planet offer a huge choice of answers, any one of which will keep you happy if what you really want is a reason to stop asking why the sky is blue. Attributing it to a deity is just a roundabout way of saying 'it just is'.

Asimov pointed out that when churches adopted lightning-conductors, they promoted science above theology. Following that way of thinking, we are trying to present scientific – or at least rational – explanations for origins, indeed for many other issues. Ponder Stibbons is the most rational of the wizards, yet even he is fighting an uphill battle. On the whole, though, he's winning, explaining Roundworld without magic, even though magic – the mechanism behind most Discworld phenomena – is his default viewpoint.

Many, perhaps most, human beings are not rational in their beliefs. Essentially they believe in magic, the supernatural. They are rational in many other respects, but they allow what they want the world to be like to cloud their judgement about what it *is* like. In the run-up to the American Presidential elections in 2012, several Republican candidates who had previously accepted basic science ended up denying it. A prominent Republican supporter opposed any kind of regulation of the markets on the grounds that this was 'interfering with God's plan for the American economy'. More extreme figures on the political right oppose taking steps to mitigate climate change – not because they think it doesn't exist, but because the quicker we wreck the planet, the sooner Christ's second coming will happen. Armageddon? Bring it on!

One reason for trying rational approaches first is that most phenomena here on Roundworld have turned out *not* to be magical. More strongly, many that used to seem magical now make a lot of sense without any appeal to the supernatural: thunder, for instance – though not the American economy, which baffles even economists. So in this book, our explanations of origins will, so far as we can manage, stick to the rational, however complicated it is. But we do wonder whether Christian Scientists, who believe it to be sinful to transplant organs, or even to transfuse blood – because they have been taught that this defies God's wishes – use lightning-conductors.

Even today, we understand less about thunderstorms than you might imagine.

Two decades ago, astronauts on the space shuttle Atlantis placed a satellite in orbit, the Compton Gamma Ray Observatory. Gamma rays are electromagnetic waves, like light, but of much higher frequency. Since the energy of a photon is proportional to its frequency, that makes them very energetic. CGRO was designed to detect gamma rays from distant neutron stars and remnants of supernovas, and it seemed clear that something was horribly wrong, because the observatory was reporting long bursts of gamma rays, emanating from … the Earth.

This was ridiculous. Gamma rays are produced when electrons and other particles are accelerated in a vacuum. Not in an atmosphere. So something was obviously going wrong with CGRO. Except – it wasn't. The observatory was functioning perfectly. Somehow, the Earth's atmosphere was generating gamma rays.

At first, these rays were thought to be generated about 80 kilometres up, well above the clouds. It had just been discovered that strange glowing lights, known as sprites and resembling huge jellyfish, existed at that height. They are thought to be an unexpected effect of lightning in thunderclouds below. At any rate, it seemed clear that sprites must be producing the gamma rays, or at least, associated with them. Theoreticians produced several explanations; the most plausible was that avalanches of electrons produced by lightning were colliding with atoms in the atmosphere, generating both the sprites and the gamma rays. The electrons could move at almost the speed of light and create a chain reaction in which each electron could kick others out of atoms.

From 1996 onward, physicists added bells and whistles to this theory, predicting the energy spectrum of the gamma rays. Data from CGRO fitted these predictions, and confirmed that the rays originated at very high altitudes. It all looked pretty good.

Until 2003.

That year, Joseph Dwyer was in Florida, on the ground, measuring x-rays from lightning, and he observed a huge burst of gamma rays from the storm clouds overhead. The burst had exactly the same energy spectrum as those that were thought to come from much higher. Even so, no one really imagined that the rays that CGRO was detecting came from thunderclouds: they were much too energetic. The energy needed to propel the rays through an atmosphere was too large to be credible.

In 2002 NASA had launched a satellite called RHESSI (Reuven Ramaty High Energy Solar Spectroscopic Imager) to observe gamma rays from the Sun. David Smith hired a student, Liliana Lopez, to look through the data for evidence of gamma rays from the Earth. There was a burst every few days, far more often than CGRO was detecting. This new instrument provided far better information about the energy spectrum, and it showed that these gamma rays had traversed a lot of atmosphere. In fact, they originated at altitudes of about 15-25 kilometres – the tops of typical thunderclouds. As new evidence piled up, it became ever harder to deny that thunderstorms generate gamma rays in huge quantities. Sprites, on the other hand, do not.

How do thunderclouds produce such energetic radiation? The answer is straight out of *Star Trek*: antimatter. When ordinary matter and antimatter come together, they annihilate each other in a burst of energy – almost total conversion of mass to energy. Antimatter powers Starfleet's vessels. Its commonest form is the positron, the anti-electron, which is naturally produced by radioactive decay and is routinely used in medical PET scanners (Positron Emission Tomography). However, naturally produced antimatter is rare, and thunderclouds are not renowned for their radioactive atoms. Nevertheless, there is strong evidence that gamma rays from thunderclouds involve positrons.

The idea is this. The electric field inside a cloud is negative at the bottom and positive at the top. This field can sometimes generate runaway electrons with high energies. Being negatively charged, these electrons are repelled by the field at the bottom of the cloud and attracted by that at the top, so they go upwards. They then hit atoms in air molecules and create gamma rays. If such a ray hits another atom, it can produce a positron-electron pair. The electron keeps going upwards, but the positron, having a positive charge, goes *downwards*, attracted by the field at the bottom of the cloud. On the way down it bumps into an air atom and knocks out new electrons … and the process repeats. Again there is a kind of chain reaction, which spreads sideways, across entire banks of storm clouds.

It's a bit like a naturally formed laser, in which cascades of photons shuttle to and fro between mirrors, triggering the production of ever more photons as they do so – until they get so energetic that they escape through one of the mirrors. The mirrors are the top and bottom of the cloud, but instead of bouncing photons to and fro, the cloud sends electrons up and positrons down. By 2005 this theory was pretty much firm. The Fermi Gamma-ray space telescope has now detected beams of charged particles, produced by thunder-clouds and travelling thousands of miles along the Earth's magnetic field lines. A substantial proportion is positrons.

This discovery puts thunderstorms in a new light. Not only is Thor's hammer creating sparks (lightning) and noise (thunder): it is creating antimatter. It's not the sort of discovery you make by trot-ting out facile explanations in terms of the supernatural. It depends on repeated scientific questioning of the known 'facts'.

Even familiar origins lead to new stories as time passes. In its search for rational explanations of origins, science often changes paradigm in response to new evidence or a new idea. The origin of the Earth and the Moon is a good example, with some curious twists. One of them being a short-term failure to change the paradigm in response to new evidence.

In this case, the main problem is too much evidence, rather than too little. We can examine the structure of the Earth, look at the record written in the rocks, and travel to the Moon and bring back specimens. But in some ways this wealth of evidence makes the problem more complicated. What does it all mean? We're trying to work out what happened, about 4.5 billion years after the event. At that time the universe had already been around for about 9 billion years (according to the Big Bang theory, and even longer according to the main alternatives). In all cosmological theories, the state of the universe gets more complicated as time passes. So by the time our solar system came into being, there was a lot of stuff around.

We have to infer, from what we can observe today, how that stuff aggregated to make the Earth/Moon system. Those observations include data from asteroids, from the Sun and the other planets, and from detailed knowledge of the structure of the Earth and the Moon. (We say 'the' Moon, but according to a recent sugges-tion perhaps there were two or more moons at one stage.) It is clear that there was a time before Earth existed, and then the Earth came into being. The Moon turned up a few hundred million years later. Their origins are intertwined, and we can't explain one while ignoring the other.

The central problem of the Moon's origin and Earth's genesis is that Moon rock is very similar, in subtle chemical detail, to Earth's mantle. This is the thick layer of rock immediately below the continental and oceanic crust, above the iron core. In particular, the proportions of different isotopes of several elements are the same in rock from either source. This coincidence is too improbable to be compatible with earlier theories of the formation of the Moon, such as the two bodies condensing independently from a primal dust cloud surrounding the Sun, or the Earth's gravitational field capturing the Moon as it was flying past. George Darwin, one of Charles Darwin's sons, suggested that the Moon was spun off from a rapidly rotating Earth, but the mechanics – such as energy and angular momentum, a measure of spin – don't work out correctly. Moreover, the Earth and Moon did not just condense from dust. Astrophysicists and geophysicists now think that the Earth aggregated from many tiny planetesimals, which formed part of a great disc with the burgeoning Sun at its centre. Our telescopes are now good enough to observe several such discs around young suns in neighbouring star systems, and many of these have been found, so that theory seems to hold up.

Between 2000 and the middle of 2012, astrophysicists and geophysicists mostly agreed that the Moon resulted from an enormous collision between an early Earth and an object about the size of Mars. They named it Theia, after the mother of the lunar goddess Selene. This collision vaporised much of the Earth, and nearly all of Theia. Most of the vapour condensed again in lunar orbit, coming together to make the Moon. The rest of it became Earth's mantle, hence the similarity. The same theory explains the large angular momentum of the Earth/Moon system, a gratifying bonus.

As time passed, problems with the Theia theory began to emerge. It would have produced very high temperatures on Earth's surface, so pretty much all of the water should have boiled away. This seems incompatible with Earth's present-day oceans. So extra assumptions were needed to save Theia. Perhaps a few ice asteroids fell on the early Earth and put the water back; perhaps the vaporised water fell back to Earth anyway. However, some very ancient Australian rocks seem to testify to the presence of a lot of water on our world about four billion years ago, which seems to be too soon after the Moon's origin for such an enormous collision to have occurred.

We described the Theia theory in *The Science of Discworld* in 1999, but by the second edition in 2002 we were no longer convinced. The biggest problem came from newer computer models of the collision. The first such models, the ones that had established the Theia theory, showed a large chunk of the Earth splashing off; then that chunk split. One part formed the Moon, and the rest fell back to Earth to form the mantle. Theia got mixed up with both of them, but in similar proportions, so anyone could see why both the Moon and the Earth's mantle had the same composition.

However, the simulations that led to this conclusion took a lot of computing time, and only a few scenarios could be explored. As computers improved, the mathematical models became more sophisticated and their implications could be worked out more quickly and more easily. It turned out that in most of them the bulk of Theia was incorporated into the Moon, while very little went into the mantle.

How can both Moon and mantle be virtually identical, then?

The proposal that was accepted until 2012 was: Theia's composition must have been very similar to that of the former Earth's mantle.

This of course is the problem that the whole theory was trying to solve. *Why should the compositions be the same?* If we can answer that for Theia by declaring 'they just were', then why not apply the same reasoning to the Moon? The Theia theory had to assume the same coincidence that it was supposed to explain.

In the second edition of *The Science of Discworld*, we described this as 'losing the plot', an opinion that Ian repeated in *Mathematics of Life*. This view seems to have been vindicated by the recent (July 2012) discovery of a similar but different scenario by Andreas Reufer and colleagues. This also involves an impactor, but now the body concerned was much larger than Theia (or Mars), and moved much faster. It was a hit-and-run sideswipe rather than a head-on collision. Most of the material that was splashed off came from the Earth, while very little came from the impactor. This new theory agrees with the angular momentum figures, and it predicts that the composition of the Moon and the mantle should be even more similar than had previously been thought. Some supporting evidence for that already exists. A new analysis of Apollo lunar rock samples by Junjun Zhang's team* has found that the ratio of isotope titanium-50 (50Ti) to isotope titanium-47 (47Ti) on the Moon is 'identical to that of the Earth within about four parts per million'.

That's not the only possible alternative. Matija Ćuk and colleagues have shown that the observed chemistry of Moon rocks could have arisen from a collision if the Earth was spinning much faster at the time – one rotation every few hours. This changes how much rock splashes off and where it comes from. Afterwards, the gravity of the Sun and Moon could have slowed the Earth's rotation down to its present 24-hour day. Robin Canup has obtained similar results using simulations in which the Earth was spinning only a little faster than it is now, but the impactor was bigger than the Mars-sized body originally suggested.

This is a case where *Pan narrans* became so committed to an appealing story that it forgot why the story was originally invented. The coincidence that it was supposed to explain faded from view, and a new narrative took over in which the coincidence took back

---

* Junjun Zhang, Nicolas Dauphas, Andrew M. Davis, Inigo Leya and Alexei Fedkin, The proto-Earth as a significant source of lunar material, *Nature Geoscience* 5 (2012) 251-255.

stage. But now the storytelling ape is rethinking the entire story – and this time it is paying proper attention to the plot.

The biggest origin question, philosophically speaking, is that of the universe, which we'll come to in chapter 18. That aside, the most puzzling origin, a much more personal one, is that of life on Earth.

How did we get here?

Our own inability to create life from scratch, or even to understand how 'it' works, makes us imagine that nature had to do something pretty remarkable to produce life. This conviction may be correct, but it could well be misplaced, because a complex world need not be comprehensible in simple terms. Life might be virtually inevitable once the mix of potential ingredients becomes sufficiently rich, without there being some special secret that can be summarised on a postcard. But explaining natural phenomena requires a convincing human-level story. That's what 'explain' means to *Pan narrans*. However, the stories scientists tell about the origin of life are generally difficult and complex, especially when it comes to filling in details. What happened probably can't be turned into a story. Even if we could go back and see what happened, what we observed might not make a great deal of sense.

Nevertheless, we can seek stories that provide useful insights.

Most scientific thinking about the origin of life considers two phases: pre-biotic and biotic. Often the problem is simplified further, to inorganic chemistry before life appeared, and organic chemistry afterwards. These are the two great branches of chemistry. The latter concerns itself with the massive and complex molecules that can be formed using lots of atoms of carbon, and the former concerns itself with everything else. And life, as it now exists on Roundworld, makes essential and ubiquitous use of organic chemistry. However, there is no good reason to imagine that the origin of life fits neatly into this convenient but rather arbitrary pair of categories. Organic molecules almost certainly existed before there were organisms to use them. So trying to understand the origin of life as some kind of sudden switch from inorganic chemistry to organic chemistry is a mistake, confusing two different distinctions.

Yes, there was a time before there was life, and a time when life was beginning. But there wasn't a sudden origin like turning a light on. There was a period, perhaps quite a long one – hundreds of millions of years – of what has been called mesobiosis. This is chemistry, both inorganic and organic, *becoming* life: the journey, not the starting point or the destination.

A large number of alternative pathways by which life might have originated have been proposed. In the 1980s Jack counted thirty-five plausible theories, and there must be hundreds by now. It is sobering to realise that we may never know which pathway actually happened. Indeed, this is quite likely. The pathway that occurred could well have been one of thousands that we haven't yet thought of. For some of us, an account which starts in chemistry and finishes in simple biochemistry is sufficient; others will want to see bacterial-grade life produced artificially in the laboratory before being convinced that the sequence of steps can work. Yet others will want to see an artificial elephant, synthesised from chemicals in bottles, and would still insist that someone cheated.

Many of you will be convinced that life is so different from the non-living, even from the freshly dead, that no account of a more or less continuous series of steps can be plausible. In part, this conviction arises from our neurophysiology: we use different areas of our brains for thinking about living or inorganic entities, mice or stones. So it is difficult for us to construct thought-chains leading from stones to mice, or even from school chemistry to 'germs'. Instead, we come up with concepts like the soul, which makes a clear distinction between our thinking about a living person, and the very different way we think about a dead body.

We'll summarise some of the plausible accounts of life's origins, so that you may enjoy the various ideas on offer and the different ways of thinking about the problem that they illustrate. We have written about the origins of life several times in the *Science of Discworld* series, so we will try to make this account a little different. The virus story at the end of this section, for instance, is quite new. It was sitting quietly behind the scenes around 2000, but in 2009 a review paper by Harald Brüssow opened it up for discussion. To put it in context, we need to look at some of the earlier proposals.

The most important early experiment was that of Stanley Miller, working in Harold Urey's laboratory in the 1950s. He imitated the effects of lightning on a reasonable approximation to Earth's early atmosphere: ammonia, carbon dioxide, methane and water vapour. Initially, he got several noxious gases, like cyanide and formaldehyde, both notable poisons; this encouraged him, because 'poison' is not an inherent property – it describes an effect on living organisms. Most gases don't get involved with life at all. Further runs of the experiment produced amino acids, some of the most important chemicals for life, because they aggregate into proteins. He came up with a variety of other small organic molecules as well.

Understanding how these molecules came into existence would be very complicated, but the experiment shows that nature can achieve the result without making any special effort. There is no reason to suppose that anything beyond standard chemistry, obeying the usual physical and chemical rules, is involved in Miller's experiments. We can tell plausible chemical tales about reasonable ways for the atoms and molecules to combine and change. It happens all the time; this is why the subject 'chemistry' exists. Reasonably detailed models would capture the main steps – but the reality is almost certainly more complicated than those models. This is an important principle: what seems complicated to us may be easy for nature.

Workers repeating this experiment with different reasonable atmospheres have obtained many other organic compounds, such as sugars, and even the bases that link up to make DNA and RNA, key molecules of terrestrial life. We've already mentioned DNA and its

135

double helix, and in any case it's very well known nowadays. RNA, which stands for 'ribonucleic acid', is less well known: it is like DNA, but simpler. With a few exceptions, RNA forms a single strand instead of two intertwined ones. Specific forms of RNA play vital roles in the development of an organism.

These two molecules could easily have been present in the early seas on our planet; indeed, they were probably unavoidable. In addition, we now know that meteorites contain many of these simple organic compounds; indeed, they can form in empty space. So that's another sensible source of organic chemicals. In short, small organic molecules were around, in quantity, for reasons that have nothing to do with living organisms.

This simple chemistry, though a promising start, isn't enough. The key molecules in organisms are far more complicated, involving vastly more atoms arranged in fairly specific ways. Graham Cairns-Smith suggested that clay molecules would be ideal catalysts for turning simple organic compounds into polymers of the kinds found in living things: linking amino acids into peptides and proteins, and possibly linking bases with phosphorus and sugars to form short strings of nucleic acids including RNA and DNA. Again, nothing beyond standard chemistry is required to achieve this, and the processes do not involve living creatures. So it would be surprising if there were *not* many polymers in the early seas. Getting complex molecules is not a problem. We may have trouble coping with their complexity, but nature just follows the rules; from this, a sort of complexity unavoidably follows.

However, polymers aren't alive. They don't reproduce, or even replicate, except in very special situations. (Replication is the making of exact copies; reproduction is the making of inexact copies which nevertheless can themselves reproduce, which is more flexible, but even harder to understand.) Replication or reproduction seem to require not just complexity, but organised complexity, and it's difficult to see where the organisation can come from. However, some of these special situations can occur entirely naturally with certain clays, which themselves exhibit replication. In a watery environment, little slabs of clay can make stacks of almost identical copies, without any help.

Since the late 1990s many things have changed. At that time, in *The Science of Discworld*, we paid particular attention to the ideas of Gunther Wächtershäuser. His proposal differed from the by then conventional Miller-Urey primeval soup, which spontaneously produced replicating nucleic acids, the first heredity. Instead, Wächtershäuser proposed that the first thing to happen was metabolism: working biochemistry. He suggested that this might have occurred where there was plenty of sulphur, iron oxide and iron sulphide, plus a suitable source of heat to drive the chemistry. One possible location that possesses these ingredients is an undersea hydrothermal vent, known as a black smoker, where molten rock from the mantle makes its way to the surface through cracks where the ocean floor is spreading. Less dramatically, underwater volcanic vents do the same. Using this kind of iron-oxygen-sulphur chemistry, Wächtershäuser came up with a set of chemical reactions that closely mimicked the Krebs cycle, a central biochemical system in nearly all living things.

In laboratory experiments, his scenario performed reasonably well, though not perfectly. So the theory of the origin of life turned into a kind of primeval pizza, with molecules dotted around on a surface, rather than a primeval soup sloshing around in pools or the open sea. In 1999 we liked this idea because it was different from heredity-first systems: we couldn't see why they would necessarily replicate – what was in it for them. Moreover, Wächtershäuser was a lawyer as well as a biochemist, and it's unusual to get good original scientific ideas from a lawyer.

However, since then a different idea, the RNA world, has really taken off. RNA and DNA are both nucleic acids, so named because they are found in the nuclei of cells. There are many other kinds of nucleic acid; some are much simpler than DNA and RNA, and some are much more complicated. Both are long chains formed by joining together four smaller molecular units, called nucleotides. Nucleotides are combinations of bases, which in turn are specific molecules that look like complicated amino acids, linked together by sugars and phosphate. Does that help? We thought not. You can look up all the details in many sources, but for present purposes we just need convenient words to keep straight which bits we're talking about.

The great trick that nucleic acids exploit is their ability to form double-chains, each half encoding the same 'information' in related ways. The DNA code letters, the four bases, come in two associated pairs, and the sequence of bases on one chain consists of the partners of the bases on the other chain. This makes possible the key feature of these pairings: each chain determines what happens in the other chain. If they split apart, and each chain acquires a new partner, by sticking on the complementary bases ... lo and behold: originally we had one double-chain, and now we have two of them, each identical to the first. The molecule not only can replicate: it does, given enough unattached bases to play with. It would be hard to stop it.

RNA has other tricks. It can function as an enzyme, a biological catalyst; it can even be the catalyst for its own replication. (A catalyst is a molecule that promotes a chemical reaction without being used up: it gets involved, helps things along and then ducks back out.) And it can also catalyse other chemical reactions that are useful to life. It's a universal fix-it molecule for living organisms. If it were possible to explain how RNA could appear in the absence of life, it would constitute a wonderfully useful step from non-living chemistry towards a primitive kind of life form. Unfortunately, it turned out to be very difficult to see how RNA could turn up in the primeval soup without any assistance. For many years, the RNA world theory was missing one of its most vital features.

This is no longer an obstacle. In recent years, many different solutions to this problem have been found, including several that work experimentally as well as theoretically. The chains of bases involved were initially fairly short – a chain of six is easy, but now there can be fifty or more bases in a chain. This is getting close to the length found in real biological enzymes, which usually have 100-250 bases. So there is real hope that long RNA chains must have been present in that ancient soup. More plausibly still, fatty membranes, which closely resemble cell membranes, have been synthesised in circumstances very similar to those that are thought to have existed on the primitive Earth, and RNA gets linked to these in useful ways. Moreover, it has recently been suggested that the RNA chains could repeatedly be broken apart – unzipped – by high temperatures in black smokers and reassembled at lower temperature in cycling convection currents. This is a lovely idea, exactly like the way DNA is multiplied in systems that analyse its sequence using the polymerase chain reaction, where alternating high and low temperatures break the chains apart and then permit them to build complementary chains, repeatedly doubling the number of copies. RNA could be replicated by this natural physico-chemical process.

For these reasons and many others, the RNA world has now become a respectable image for the earliest stages of life on Earth. It may not be what actually happened, but it provides a plausible scenario. And even if life did not arise in that way, the RNA world shows that there is no compulsion to invoke the supernatural. In the primitive seas, probably around smokers but perhaps on tidal beaches where pools were concentrated – and irradiated – in sunlight, and diluted when the tide came in, or under the influence of volcanoes or earthquakes, RNA strings were growing and being copied.

The copying process wasn't always perfectly accurate, but that was a positive advantage, because it led, without any special interference, to diversity. If random variation of this kind could be coupled to some kind of selection, favouring sequences with specific features, then RNA could – had to – evolve. And selection wasn't an issue; the big problem would have been to prevent it. As special sequences with particular properties appeared, competition between them for free nucleotides, and probably for interactions with particular fatty membranes, eliminated some sequences and encouraged others to proliferate. This led to longer chains with even more special properties.

Natural selection had begun ... and the system was *becoming* alive.

In this view, not only does evolution by natural selection explain how life diversified: it is part and parcel of what brought it into being in the first place. Copying errors, if they occur, though not *too* often, can be creative, in the context of a sufficiently rich system.

The RNA world is not the only game in town. The latest proposals for the origin of life hinge upon viruses. Viruses are long DNA or RNA chains, usually surrounded by a protein coat that contrives to inject them into other organisms, especially bacteria and animal and plant cells. Most viruses rely on the DNA/RNA copying machinery of the organism they infect to replicate them. Then the new copies are usually sprayed out into the environment when that cell, or the organism, dies.

Since the work of Carl Woese in 1977, taxonomists – scientists who classify life into its innumerable related forms – have recognised three basic kinds of life form, the largest and earliest branches of the tree of life. These 'domains' comprise bacteria, archaea and eukaryotes. Creatures in the first two domains are superficially similar, being micro-organisms, but each domain had a very distinctive evolutionary history. Archaea may trace back to the earliest organisms of them all; many live in strange and unusual environments: very hot, very cold, lots of salt. Bacteria, you know about. Both types of organism are prokaryotes, meaning that their cells do not clump their genetic material together inside a nucleus, but string it from the cell wall, or let it float around as closed loops called plasmids.

The third domain, eukaryotes, is characterised by having cells with nuclei. It includes all complex 'multicellular' creatures, from insects and worms up to elephants and whales. And humans, of course. It also includes many single-celled organisms. RNA sequences imply that the first big split in the tree of life occurred when bacteria branched away from ancestral archaea. Then that branch split into archaea and eukaryotes. So we are more closely related to archaea than we are to bacteria.

Viruses are not part of that scheme, and it is controversial whether they are a form of life because most of them can't reproduce unaided. It used to be thought that viruses had two different origins. Some were wild genes that escaped from their genomes and made a living by parasitising other creatures and hijacking their gene-copying equipment. The others were desperately reduced bacteria or archaea; in fact, they were reduced so far in their pursuit of a parasitic lifestyle that all they had left were their genes. Occasionally it was supposed by lay people, physicists or maverick biologists – who should have known better – that, being so simple, they were relics of ancient life. This incorrect line of thinking stems from the same, mistaken, principle as the one that considers *Amoeba* to be ancestral because it looks simple. Actually, there are many kinds of amoeba, and some have 240 chromosomes, bodies in the cell that contain the genes. We have a mere 46 chromosomes. So in that sense an amoeba is more complex than we are. Why so many? Because it takes a lot of organisation, in a very small space, to get all of an amoeba's functions to work.

Brüssow, in 2009, wrote a review called 'The not so universal tree of life *or* the place of viruses in the living world'.* In it he pointed out that Darwin's tree of life, a beautiful idea that derives from a sketch in *The Origin of Species* and has become iconic, gets very scrambled around its roots because of a process called horizontal gene transfer. Bacteria, archaea and viruses swap genes with gay abandon, and they can also insert them into the genomes of higher animals, or cut them out. So a gene in one type of bacterium might have come from

---

* Harald Brüssow, The not so universal tree of life *or* the place of viruses in the living world, *Philosophical Transactions of the Royal Society of London* B **364** (2009) 2263-2274.

another type of bacterium altogether, or from an archaean, or even from an animal or a plant.

The major agents of this swapping are viruses, and there are lots of them on the planet, probably ten times as many virus particles as all other forms of life added together. Now, it might appear that with all this swapping, it would be virtually impossible to work out lineages, the heredity of individual bacteria. Even more so, it would seem impossible to work out the lineages of the viruses doing the swapping. Surprisingly, this is not the case; well, not altogether. There are clues in the order in which specific genes appear in many viruses, and there are useful clues as to which organisms the viruses parasitise. Some parasitise both bacteria and archaeans, suggesting strongly that they have been doing so since before these groups diverged. Moreover, these are viruses with RNA genetics. So Brüssow proposes quite convincingly that these particular viruses may be relics of a former RNA world. Going further: infection by ancient DNA viruses could have imported DNA into the heredity of all of the familiar creatures whose genomes we now make such a fuss about. So some of the mavericks, and physicists, might have been right all along, even if it was for the wrong reasons.

If that is the case, we need to look with new eyes at all the ways in which RNA is involved in modern life forms. According to the standard story, which has not changed for some time, RNA serves as a humble messenger that carries the all-important DNA sequence to the ribosomes, huge molecular structures in which proteins are assembled. There are also small RNAs that transfer amino acids to the ribosomes for protein assembly. Ribosomes in turn are made of several kinds of RNA, and several people have suggested that they are the central mechanism of protein construction in cells, their most important function.

This story may soon have to change.

There has been a revolution in nucleic acid biology over the last ten years, and almost all of it has been about RNA. Messenger

RNA and transfer RNA are merely the most prosaic jobs that RNA performs in cells. But RNA does many more important – perhaps having said 'prosaic' we should now say 'poetic' – jobs too. When DNA was considered the most important molecule in the cell, and protein construction the most important function (many textbooks still think so), strings of DNA that specified proteins by transcribing messenger RNA were called *genes*. The strings of DNA upstream and downstream, which did not specify proteins, were mostly thought to be 'junk DNA' of no importance to the organism. Junk DNA was just there as an accidental by-product of past history, but because it didn't cost much to replicate it, there was no evolutionary reason to eliminate it.

Indeed, there are plenty of remnants of old genes, and quite a lot of sequences from ancient viral attacks, which really might be junk. However, it turns out that although it doesn't specify proteins, nearly all DNA between genes does transcribe RNA molecules. These RNA molecules form the main control system of cells: they control which genes are activated and when, and how long different messenger RNAs last before destruction. In bacteria they also control genes, but a subset of them protects the bacterial cell against attack by viruses. This is a simple kind of immune system. So DNA may be the soloist, but RNA is the orchestra.

With that established, we can return to ribosomes, the molecular factories that assemble proteins. They are tiny particles, mostly of RNA. In bacteria, archaea, animals, plants and fungi, every cell has its own complement of ribosomes; moreover, much the same RNA occurs, though with different proteins, right across the span of life.

Marcello Barbieri is a leading exponent of biosemiotics, a relatively new science concerned with the *codes* of life. You have probably heard of the genetic code, the way in which triplets of DNA nucleotides are turned into different amino acids in proteins – by the ribosome. Barbieri has pointed out that there are hundreds of other codes; they range from insulin anchoring itself to receptors on the cell surface and causing different effects in the cell, to a smell (technically a pheromone) in male mouse urine that affects the oestrous cycle of female mice. All such effects are the results of translating one chemical language – hormones, pheromones – into a different language – physiological effects. So the genetic code is not alone: there are codes everywhere in biology. From this viewpoint, the crucial element in protein formation is not the DNA that prescribes it, or the messenger RNA that transmits the prescription: it's the ribosome. Which, to complete the analogy, is the pharmacist that makes up the prescription.

It also seems clear that this very ancient piece of machinery, so central to all living function, pre-dates the bacteria/archaea split, so it probably derives fairly directly from the RNA world. Something back then formed a relationship, a translation, probably from nucleic acid to protein. The ancestor of today's ribosomes, probably not very different from today's range of RNA structures, did the trick. So at the beginning of life, we find the translation of one kind of chemistry into another, by a structure that has come down to us almost unchanged.

Before the ribosome, there was just chemistry. Complicated chemistry, to be sure, but complication alone isn't quite the point. What matters is complexity, which in this context means 'organised complication'. Every cook knows that heating sugar with fats, two fairly simple chemical substances, produces caramel. Caramel is enormously complicated on a chemical level. It includes innumerable different molecules, each of which has thousands of atoms. The molecular structure of caramel is far more complicated than most of the molecules you're using to read this page. But caramel doesn't do much, aside from tasting good, so mere complication isn't enough to make interesting new things happen. Similarly, mixing dilute solutions of amino acids, sugars, bases and so on with particular clays generates long, very complicated, polymers. But, like caramel, they're not very interesting. However, as soon as transactions between those molecules came about, via the earliest ribosomes, complexity took over from complication.

Here, 'complexity' refers to *organised* complication. In a complicated system, such as a car, the individual bits – brakes, steering wheel, engine – behave in much the same way outside the system as they do when they're part of it. Mostly, they just sit there unless they're pushed, or pulled, or operated, by something else. But you, a fly, or an amoeba, are different. Their components behave differently when they are part of the system compared to what they do on their own. The parts interact more closely, changing their nature in the context of the system.

A bridge linking an island to the mainland is a complex system in this sense. In order to do its job, it doesn't much matter what the bridge is made of: it could be rope, steel or concrete. It could even be made of nothing (or air) if it's a tunnel. The important property is not what it's made from, but that it links the two ends effectively. That linkage is an *emergent property* of the bridge. That is, it's not inherent in any of the materials used. It arises because of their relationships to each other and to the local geography. Moreover, once the bridge is in place, the local geographical function is changed. The river that the bridge spans is no longer an obstacle to vehicles, even though they can't float or travel underwater. Crucially, you won't understand how that change occurs by studying the materials that made the bridge.

When the two ends of the bridge are linked, and only then, the local geography changes dramatically. So the real origin of a bridge occurs *when the ends are linked.* For some purposes, this is when the first rope crosses the divide; for other purposes it's when the first car makes the crossing; for yet other purposes it's when the Customs Office is set up.

Similarly, a ribosome in a cell is very different from an isolated one. It has a specific but complex job to do, reading messenger RNA and constructing proteins according to the genetic code. We wonder whether the chemical transactions made possible by early ribosomes in effect constructed bridges between several different kinds of chemistry, providing energy and materials for the ribosome to replicate itself. It's mostly RNA, after all.

Indeed, if we had to point to a single innovation that marked biotic from pre-biotic, it would be the ribosome, the translator supreme. Barbieri thinks the ribosome is central to life, and so do we. DNA is simply the rather prosaic, boring text. The ribosome is the orator; the other RNAs are the poetry. Once the ribosome emerged, the future became a living future, and in many ways this step marks the true origin of life.

Most origins also involve more subtle forms of emergence: the beginning of a storm, the acorn's origin as a bud on the oak, the origin of the Earth. Each of these origins is a quantitative-to-qualitative transformation, an emergent event that localises a real beginning. The first stroke of lightning, the first pair of leaves, the generation of heat that melts the core inside the Earth's mantle: these are emergent events that can label beginnings of new structures. The 'becoming' has divided into two issues, before and after the emergence.

If a phenomenon is emergent, it transcends all that has gone before. It does something that its bits and pieces could not have done on their own, or partially assembled, or assembled with some extra scaffolding that gets in the way. This transition is often the best stab we can make at assigning an origin. An emergent phenomenon does not originate in the bits and pieces that led to it: it originates when it emerges.

The emergence of the first lightning strike marks the beginning of the storm. The cell divisions that mark the acorn's difference from the other buds around it are the emergent oak. The cell divisions and relationships that promoted the egg that later became you orchestrated the emergent event that began you. The universe is complicated because emergent events – quantitative differences becoming qualitative differences – have occurred so many times.

146

Bridges like ribosomes have been built, and the Moon now circles the Earth.

These links have joined separate events into a web of causality that is the most notable property of the world around us. A story, however, is not a web. It has a linear structure, because both speech and writing proceed one word at a time. Even hypertext, used on the internet, is determined by a linear programme written in hypertext mark-up language (html). And that is why storytelling – human narrativium – finds origins to be so difficult and puzzling, and sometimes looks for simplicities where none exist.

Evolved, or designed?

Maybe the difference isn't as great as most of us think.

When design is presented as an alternative to evolution, there is an unstated assumption that the two are very different. Design is a conscious process carried out by a designer who knows what result he, she, or it is aiming at, and whose purpose is to achieve it. Evolution selects, from a lot of random variants, changes that lead to some kind of improvement in survival prospects; then it makes lots of copies of the successes. It has no aims and no purpose. It is not 'blind chance', a description that creationists often use, forgetting (we are being charitable) the crucial element of selection. But the process of evolution is exploratory, not goal-seeking.

On closer examination, however, design and evolution are much more similar than most of us imagine. Technology appears to be designed, but mostly it evolves. Improved technology is selected because it works better, and it then displaces earlier technology. This process is analogous to the way that natural selection causes organisms to evolve, so it is reasonable to speak of technology evolving. (The analogy is loose and should not be taken to extremes. Technical drawings or CAD designs are poor analogues of genes.) Selection of technology may appear to be by human agency, but it is highly constrained. Success is decided by vote, and voters vote with their wallets. The inventor's intentions are almost irrelevant. Just as in biological evolution, the main constraint is *what works*.

Because of the difficulties inherent in the simple-minded approach to design that Paley proposed – because designers don't go straight from idea to design to product – we should look carefully at just how designs appear in technology. That changes our attitude to 'design' in nature too.

Most human designs don't work the first time round. Most jugs still don't pour well. It's cheaper to invent a new type of jug, even if it's no good, than it is to pay licensing fees for one that works. The better mousetrap, even when it is genuinely better, is a minor variation on hundreds of previous mousetraps. Usually.

Mousetrap evolution is a coherent process, not just a succession of unrelated gadgets. The same goes for bicycles, cars, computers, even jugs. Each new capability causes a particular technological path to branch, leading to new roads. Stuart Kauffman, one of the founders of complexity science, introduced the term 'the adjacent possible' to mean the possible behaviours of a complex system that are just a short step away from wherever it currently is. The adjacent possible is a list of what potentially might develop. In a sense, it *is* the system's potential.

Organic evolution proceeds by invading the adjacent possible. Invasions that fail aren't invasions at all, and nothing much changes. Successful invasions don't just change the system that does the invading; they change the adjacent possible of *everything*. When insects first took to the air, the ones that stayed on the ground were suddenly in danger of predation from above, even though they hadn't changed at all. Likewise, technology advances by continually invading the adjacent possible. Technological evolution is faster than organic evolution because human minds can use their imaginations to jump into the adjacent possible and see if it works, without actually doing it. They can also copy, which organic evolution does only rarely, aside from reproducing near copies of organisms. These are processes that generate paths and histories, and contexts in which some evolutionary trajectories are viable, but others are not. Only a few select trajectories work. In contrast, thinking in terms of innovations that generate products makes the design process work like magic.

There are a few useful analogies between technological evolution and organic evolution, and a lot of misleading ones. Comparisons between organic evolution and economics abound in the literature, and nearly all of them are misleading, from social Darwinism to the 'cost' of reproduction. Some evolutionary trajectories, however, can usefully be compared to biological ones. Examples include telegraph

→ telephone, especially international with undersea cables as investment, pens → word processors, and rockets → space elevators, which we'll come to shortly. These changes get rid of old constraints at each subsequent recursion.

There are biological precedents, in which evolution did *not* lead to increased complexity (as measured by DNA information), but the reverse. One is the evolution of mammals. Mammals have *less* DNA than their 'more primitive' amphibian ancestors, a trick that can be pulled off because mother mammals control the temperature of their developing embryos by keeping them inside their own bodies. Amphibians need huge quantities of genetic instructions to plan for many different contingencies, as their embryos grow in a pond, subject to the unpredictable vagaries of the weather. Mammals dispense with this excess baggage by investing in temperature control.

With the expanding possibilities of the chemical/physical universe as a substrate, and organic evolution as a model for an emergent phase space, we should be asking 'What are the constraints on technology, if any?' rather than 'What is the pattern of technological advance?' Sometimes there are persistent patterns. Moore's Law states that computing power doubles every eighteen months. It has worked for decades, even though (indeed, because) technologies have changed dramatically. Some experts think that the increase in power will shortly have to slow down, but others remain convinced that new ideas, often already visible, will keep it going.

Our culture sometimes seems to follow evolutionary trajectories too. As individuals, we respond to the cultures in which we live, and we are conducted into our technological future as it changes progressively. As far as cultures go, this is an evolutionary process. From a human viewpoint, however, such progressive change looks like the development of a more complex living system, a socio-dynamic. Is technology cancerous, born of mutation as it burst out of its hunter-gatherer background, as it evolves into new forms? Or is it developmental, exploiting new organisations as it invents them but maintaining an adaptable but stable path, just like a developing embryo? An embryo destroys many organised structures, and kills many of its cells, as it develops. It builds scaffolding and throws it away when it is no longer needed.

From the point of view of the individual human, caught up in a technological rat race, this stress is clearly a symptom of social pathology, as Alvin Toffler argued in *Future Shock*. In contrast, looked at culturally, it is natural development. This difference of viewpoint resembles the two ways to describe a thinking mind: nerve cells and consciousness. More generally, not only can every complex system be described in several non-overlapping ways: it can also be described on several levels ... as concrete or as a bridge; as an architectural bridge or as a weak point for an enemy invasion.

Human evolution occurs on two levels: embryological development and cultural development. Neither process is preformational, with all necessary ingredients already present. Neither is a straightforward blueprint: make it like *this*. In both, evolutionary changes occur through complicity between several programmes, each of which affects the future of the others. As time passes, each programme not only affects its own future by its own internal dynamics: it also changes its future by the changes it causes in the other programmes.

To what extent are those changes predictable or accidental? There is a difference here between two modern viewpoints, one associated with the palaeontologist Simon Conway Morris in *Life's Solution: Inevitable Humans in a Lonely Universe*, and the other with the late Stephen Jay Gould in *Wonderful Life*. This difference is crucial to the issue of design in evolution.

Gould made great play of the variety of the animals represented by the fossils in the Burgess Shale, deposited at the start of the Cambrian period about 570 million years ago. These fossils had been described by previous biologists, but Morris reworked and reconstructed them. He classified these fossils into a wide range of morphological types; in fact, many more basic kinds of animal design ('phyla') than had been assigned previously. Gould used this wide range of body designs, only a few of which have descendants among present-day creatures, to argue that life can do almost anything by way of morphology, even in its fundamental or basic structure, and that the organisms that now exist are accidental survivals from the much vaster range that existed at the start of the Cambrian.

Morris, however, has come to believe the opposite, namely: because some of the many themes have converged to produce similar beasts, some specific designs must be winners, no matter how they are realised. Therefore any wide array of different body structures will necessarily evolve to generate much the same spectrum that we observe today, automatically selected because those are the body-plans that work best. The fossil record contains many cases of this kind of convergence:* ichthyosaurs and dolphins have evolved to look like sharks and other carnivorous fishes, because that's the shape that's most efficient for a fishy predator. In short, Morris believes that if we were to find living creatures on a similar planet to Earth, or if we were to run Earthly evolution again, then much the same range of animal designs would appear. Aliens on a world like ours would be much like us, even if their biochemistry were totally different.

In contrast, Gould believed, as we do,† that in such a rerun the resulting spectrum of life forms would not resemble the current ones at all. Different designs, fundamentally different body forms, would be just as likely as the ones that happen to exist now. The current body-plans are just a contingent, accidental collection that happened to survive. Aliens, even the highest ones, would most likely be very different in design from us, whatever world they evolved on. Including a reboot of ours.

---

* Jack recalls a bright Irish student who, in an exam question about convergent evolution, defined it as 'where the organs of two descendants are more alike than they were in the common ancestor'.
† See Jack Cohen and Ian Stewart, *What Does a Martian Look Like?*

The old view of the role of genes in Darwinian evolution emphasised mutations: random changes to DNA sequences. However, at least in sexual species, the main source of genetic variability is actually recombination: mix-and-match shuffling of gene variants from the parents. New mutations are not needed to innovate; new combinations of existing genes are sufficient. The diversity of available gene variations can be traced back to much older mutations, but you don't need a mutation *now* to change an organism.

All biologists now agree that the body-plans of organisms are not built up piece by piece, mutation by mutation, but have been selected by recombination. Instead of mutations to new genetic variants, we find recombinations of many ancient mutations. These are sorted from kits of compatible parts in every generation, not put together higgledy-piggledy and expected to work. If, as seems plausible, only a few developmental trajectories can lead to larvae that can feed and grow into working adults, compared to the huge number that can't, then it is to be expected that the successful designs are all separated, without intermediate forms bridging the gaps. 'Missing links' need not be missing – or links – because continuous variation is not required in a discontinuous process.

By looking at so-called *r*-strategists, animals like plaice and oysters whose larvae comprise only a few developmentally competent ones among a majority that aren't, we can see how this is achieved today. What it does not tell us – what distinguishes the Morris and Gould views – is whether the successful designs are out there in some Platonic organism-space, waiting to be found, or whether the organisms have all invented their own, unpredictable, designs as they went along. Morris, a Christian, believes the former: the appearance of design is the revelation of transcendental attractors in God's design-space of possible organisms. We, however, believe that there are so many possible ways of being a successful organism, so many effective designs, that the drunkard's walk of evolution keeps finding them, even though they are sparsely embedded in the vast majority of failures.

In particular, we think that intelligent design focuses too narrowly on the evolution of *specific* structures found today, such as the precise molecular configuration of haemoglobin or the bacterial motor. In retrospect, these structures seem highly improbable; if nature were to aim for them again, it would almost certainly miss. But evolution selected these structures when it encountered them. What matters is how likely it is that evolution could find *some* such structure, not that *specific* one. If there are many suitable structures, then a process that automatically homes in on anything that seems to be an improvement has a good chance of finding one of them.

Think how improbable *you* are. If two genomes had not combined just so, if that egg and that sperm had not come together, if your father hadn't met your mother at the dance, if the wartime bomb in the harbour had hit your grandfather instead of being a hundred metres away, if Napoleon had won the Battle of Waterloo, or if victory had gone the other way in the American War of Independence, if the nascent Earth had not acquired an ocean, or the ripples in the Big Bang had been slightly different … you wouldn't be here.

The odds that you exist are infinitesimal.

No. The odds that you exist are certain, *because you do.*

The processes that led to you are robust, and at each stage would have led to something similar, albeit different, if run again. No complex process ever produces the same result twice. But if it produces a similar result instead, that makes its consequences certain, not utterly unlikely. Only fine details will be different, second time around. The lottery of life is quite different when seen through the eyes of the eventual winners, rather than those of a random competitor before the contest has happened.

It's tempting to assume that the evolution of technology can tell us about organic evolution, or vice versa, but these processes lead to apparent design in very different ways. However, there is a grand overall similarity in how we *think* about both systems, particularly how our thinking has changed over the last few years. The appearance of design is the most dramatic element in both systems. Although its provenance is different in the two cases, we are no longer surprised by it. We have realised that the universe is *not* doomed by increasing entropy to an eventual 'heat death', a traditional but somewhat misleading term which actually means that the universe will end up as a structureless lukewarm soup. Instead, the universe 'makes it up as it goes along', and what it makes up are designs. In that sense, at least, the appearance of new design in both technical and organic systems can be considered comparable. But it's important not to stretch the metaphor too far.

Cultures can also be seen as evolving. In many ways, cultural evolution sits between organic and technological evolution. Advanced human societies make their members different and varied. All societies produce numerous distinct roles, from those limited by sex and age, such as childbearing or going to school, to those that seem to be chosen by the individual: warrior, accountant, thief. There is a division among sociologists that is comparable to that between Morris and Gould. Some believe that the roles are in some sense transcendent or universal; they look for proto-accountants in 'primitive' hunter-gatherer societies. Carl Gustav Jung's theory of archetypes, such as the persona, the shadow, and the self. In his view, these were extremely ancient common images derived from humanity's collective subconscious, which affected how we interpret the world. Others, however, believe that some roles in different societies, even though they look similar and the names translate similarly, can be fundamentally different: a Japanese car worker has a different worldview from that of his English equivalent, and occupies a different societal slot.

Both sociological viewpoints can provide useful insights: different societies, like different ecologies and different cultures, provide diverse roles for their members. The cultural invention of generic occupations is comparable to the organic invention of things like chordates, trilobites, muscles and nests. It is also comparable to the technological inventions of – say – bicycles, the internal combustion engine, wheat and rope. Money in human societies is usefully analogous to the way cells produce and exchange energy, using the molecules ADP and ATP (adenosine di- and tri-phosphate). Indeed, ATP is often called the unit of molecular currency. The appearance of new designs in organic evolution, in cultures, in technology, and even in language, can usefully be compared. Even so, such comparisons must be made very carefully and not pushed beyond reasonable limits.

The idea that technology evolves is not the orthodox view, wherein design and evolution are considered to be opposites. Design in technology is usually seen as being invented, not as having evolved. This assumption lies at the core of Paley's famous analogy between a living creature and a watch. Watches are intricate devices, designed and made by an intelligent agency. Therefore, if you find something equally intricate in living creatures, it must also be designed, and the creatures must have been made by an intelligent agency. Therefore there must have been a cosmic designer, QED. The same assumption motivates the current hypothesis of intelligent design, which is basically Paley's argument restated using examples from modern biochemistry.

However, analysis of the history of nearly all inventions shows them either to be developments of previous technology, that is to say adaptations, or perversions of some technology in a different sphere. (A few do seem to come out of thin air, with no significant precursors.) The biological term for such things is 'exaptations', a word introduced by Gould and Elizabet Vrba in the 1980s. It refers to an organic or a technological development that arises from an entirely different structure or function. An example is the use of feathers for flight. Feathers first appeared in dinosaurs, but their skeletal structure shows that the early feathered dinosaurs didn't use their feathers to fly. We can't be certain what they did use them for, but the most plausible functions are for warmth or for sexual display. It may well have been both. Later, feathers turned out to be useful for wings and flight, and birds evolved. Nature is an opportunist. A technological example of exaptation is the use of disc-recorded sound for music. Edison originally developed the phonograph for a more serious purpose, to record for posterity the last words of famous men and the historical speeches of politicians. He greatly deplored its use for frivolities like music, but accepted payments gracefully, nevertheless.

Exaptation is one of the less obvious tricks that evolution has up its sleeve, and is often the solution to evolutionary puzzles, in which a particular function can occur only when several interrelated structures apparently have to appear simultaneously, but none of them can perform that function on its own. Although it's tempting to deduce that such structures can't evolve at all, they can if exaptations occur. Then the structures concerned initially perform different functions.

A classic instance is the bacterial flagellum, a structure that proponents of intelligent design argue cannot possibly have evolved by any conceivable route. The flagellum allows some bacteria to move of their own volition. Its most important component is a tiny molecular motor, which causes the flagellum to rotate, much as the motor of a boat turns the propeller. The bacterial motor* is made from a large number of different protein molecules. Until recently, evolutionary biologists could offer no convincing explanation of the origin of such a complex structure by natural selection.

In 1978 Robert MacNab wrote: 'One can only marvel at the intricacy, in a simple bacterium, of the total motor and sensory system ... What advantage could derive ... from a "preflagellum" [meaning a

---

* We say 'the' motor, because everyone does, but different bacteria have different motors. Darwin was puzzled why the deity would design hundreds of very similar barnacles, all of different species; we may similarly wonder why an intelligent designer would intervene in the normal process of evolution to equip dozens of bacteria with individually designed motors.

subset of its components], and yet what is the probability of "simultaneous" development?' In 1996 Michael Behe, a biochemist and leading proponent of intelligent design, repeated MacNab's worries in *Darwin's Black Box*, together with several similar evolutionary puzzles. He concluded that while many, indeed most, features of living creatures have evolved, some cannot possibly have done so because they are irreducibly complex: if you remove any component, they cease to function.

It's a genuine puzzle, but before invoking some unspecified genie-of-the-lamp, without independent evidence that it exists, we ought to make sure that conventional evolutionary processes definitely can't hack it. Intelligent design doesn't just argue that some specific evolutionary route is wrong: it claims a proof that *in principle* no such route can exist. If you're going to invoke a general principle of this kind to assert the existence of a supernatural being or a highly advanced cosmic designer, you need to close any loopholes in your logic. Otherwise your entire philosophy will be built on sand, whatever actually happened. The Book of Genesis could be true in every detail, but your supposed proof would still be nonsense if its logic were defective.

In response to intelligent design, biochemists have taken a closer look at the proteins in the bacterial motor and the associated genes. The most prominent components of these motors are rings of proteins, which are very common in evolution. What use is a ring? It has a hole. Holes are amazingly useful to a bacterium or a cell, because they can function as pores or sockets. Pores let in molecules from the outside world, or expel molecules into the outside world. Different-sized pores deal with different-sized molecules. That's something that natural selection can work on: a mutation in the DNA that codes for the protein can lead to one with a similar, but slightly different, shape or size. As soon as a pore does a useful job, evolution can find a pore that is better at doing that job, if there is one.

Sockets allow bacteria or cells to attach new structures, either inside or outside the cell membrane. Many different molecules can fit into the same socket, and again, evolution has plenty of opportunities to work with. What began as a pore can become a socket if something happens to fit into it. When the two modules come together, their function may change. Exaptation demolishes irreducible complexity as an obstacle to evolution. You don't even have to prove exactly how a given structure evolved, because irreducible complexity allegedly rules out not just the actual route, but *any conceivable one.*

So let's conceive.

A number of biologists have attempted to deduce a plausible or likely evolutionary route to a bacterial motor, from DNA and other biochemical evidence. This turns out not to be especially difficult. Many details are still provisional, as is all science, but the story is now sufficiently complete to disprove the contention that the motor exhibits a type of complexity that rules out *all* evolutionary explanations. Agreed, that doesn't prove that the current evolutionary explanation is correct. That must be confirmed, or denied, by further scientific investigations. But it's quite different from asking whether, in principle, any such explanation can exist.

The most fully developed synthesis of these proposals, put together by Nicholas Matzke, starts with a general-purpose pore. This evolves into a pore with more specific functions. At this early stage, the structure is not a motor, but it already has a very useful, entirely different, function: it can transport molecules out of the cell. In fact, it is recognisable as a primitive version of so-called Type III Export Apparatus, which exists in modern bacteria, and DNA sequences support this. Further changes, in which the pore's function is successively improved, or changed by exaptation, provide an entirely plausible evolutionary route to the bacterial motor, increasingly supported by DNA evidence.*

* N.J. Matzke, Evolution in (Brownian) space: a model for the origin of the bacterial flagellum, www.talkdesign.org/faqs/flagellum.html.

Yes, if you take away enough parts of the bacterial motor, then it might not be a very good motor any more. But evolution didn't know it was supposed to be making a motor.

So 'design' isn't what it is often thought to be, even for human technology, let alone biology. Each innovative step may be driven by human intentions, but what works, and what passes on to later technology, evolves. To some extent, cars evolved from horse-drawn carriages, and a ballpoint pen is the lineal descendant of a quill made from a feather. We can legitimately compare these developments to mammals evolving from a Devonian fish that came out of the water onto land, or to our little middle-ear bones being the lineal descendants of bony gill structures in that fish.

Evolution is not efficient. It throws an awful lot of things away. Innumerable land vertebrate species have gone extinct. Similarly, most human designs don't work. From the enormous number on offer, only a few develop into sophisticated structure/function niches. We are all bound by tradition, as well as by functional constraints that require any new development to fulfil the same functions as its ancestor. There's a classic example: Apollo rockets were moved to their launching-pads on rails that were much too close together for stability, because the gauge of America's railways came from mine railways that were two horses wide. So the Moon project was jeopardised by horse's asses.

To be specific, let's think about better mousetraps. Mousetrap evolution is a process, not just a succession of models; it branches into the future. The pattern that has a metal bar coming down and (one hopes) breaking the mouse's neck, has expanded into dozens of different models, some computer-controlled. Those that trap the mouse in a metal tube, or a cage, are more like descendants of lobster-pots, but these too have performed what biologists would call an adaptive radiation: we found seven different kinds, with sprung doors or elastic apertures for entry.

The same goes for bicycles, cars or computers: they all adaptively radiate into the future. Each new ability, such as computer control – a logic chip – on a particular technical road branches into new roads. Think of the familiar cat flap, now available in versions that allow your own cat, wearing its magnetic collar, in or out, but exclude foreign cats. Or fancy electronic ones that verify your cat's ID. Full-body scanners to detect terrorist cats carrying exploding mice cannot be far away. Just as in organic evolution, the adjacent possible is continually being invaded: possibilities just one step away from current practice are tried, rather unoriginally.

We usually think of this as technical development, not innovation, unless it is in an unexpected direction: Teflon used for non-stick frying-pans, or penguins' wings used for swimming. Most aquatic vertebrates, unlike these birds that have become secondarily aquatic, use their tails, not their fins, for propulsion. Such more original changes of direction are best thought of as exaptations rather than adaptations. Or, to use a less biological term, genuine innovations.

Among those who accept evolution as a reasonable metaphor for many examples of progress in technology, it used to be thought that the major difference between technical and organic evolution is that technological evolution is Lamarckian – named after the French naturalist Jean-Baptiste Lamarck, a contemporary of Darwin – whereas organic evolution is Darwinian. In Lamarckian evolution, acquired characteristics can be inherited – if a blacksmith acquires strong arms because of his work, his sons should also have strong arms. In Darwinian evolution, that's not possible. Neo-Darwinism illuminates the difference: heritable characteristics are those that are determined by genes.

Lately, this distinction has become a bit blurred, and each mechanism has acquired features that were thought to be characteristic of the other. Technical development has borrowed a trick from evolution to construct so-called genetic algorithms for the development of new products. Digitised designs are shuffled, by analogy with recombination, the way biological reproduction shuffles gene variants from both parents. The next technological generation to survive this process combines the more useful features of previous generations. Sometimes it has new emergent properties, which are selected if they prove useful, and are retained. Often the final design is incomprehensible to a human engineer. Evolution need not obey human narrativium.

The phenomenon of genetic assimilation, which is entirely Darwinian, can look very Lamarckian. Changing a population progressively by selecting genetic combinations that work can change the thresholds at which particular capabilities come into play. As a result, effects that originally depended on some environmental stimulus can happen without that stimulus in later generations. For example, the skin on the soles of our feet gets thicker when we walk regularly, an acquired characteristic; however, genetic recombinations that provide babies with thicker skin on their feet from the start make this process more effective, and so are selected for. Any new feature, acquired or not, that *works* – that improves the chances of surviving to reproduce – reveals a feature that Darwinian evolution could blunder into and exploit. Genetic assimilation may indeed be the usual way that originally responsive adaptations get built in to the developmental schedule.

In particular, the old distinction between Lamarck and Darwin has lost its power to distinguish technical from organic evolution. But that doesn't imply that there are no significant differences. It's tempting to think that one obvious aspect of technological evolution surely can't apply to Darwinian evolution: *imagining* a possibility before designing a technique or gadget to implement it. Human technology is born in the imagination of a series of inventors or discoverers: 'What would happen if … ?' is a theoretical exploration of Kauffman's adjacent possible. Much of the time, imagining possibilities leads to hypothetical new inventions being rejected without bothering to make them or test them: they wouldn't work because ... *or* no one could use them because ... *or* they would be too expensive ... *or* they wouldn't perform well enough to displace the widget that already does the job very well.

It doesn't seem possible that this imaginative process could have an organic analogue – but it does. In 1896 the psychologist James Mark Baldwin wondered whether animals carrying out behavioural experiments might be drawn into the evolutionary process, in effect by imagining what would happen if they could perform some new task that was actually beyond them. For instance, an okapi is like a giraffe, but its neck and legs are of normal length. Suppose that an adventurous okapi, for example, kept reaching up in an attempt to browse on the lowest branches of trees, despite repeated failure. *Because* it failed, this would be analogous to imagination. But occasional success could favour okapi with slightly longer necks and legs, leading to a giraffe. This process is often called the Baldwin effect.

A few years ago, we observed some animal behaviour that could well become the root of such an evolutionary trajectory – an exaptation in the making. Plecostomid catfish ('plecs') are common scavengers in larger aquariums, cleaning algae off the glass with their sucker-like mouths. In the wild, they can hold tight to smooth rocks as they glean the algal film; they also have effective armour with barbed bony supports in their dorsal and pectoral fins. In aquaria, these characteristics give them an entirely different ability, which we saw a plec exploiting in the Mathematics Institute common room at the University of Warwick. This plec's natural abilities made it much better than other fish at garnering floating pellets of food. It did this using a method quite alien to wild plecs: it turned on its back and used its sucker-mouth to take in soggy pellets, while its spiky fins kept off the competition. So a catfish mouth, adapted for taking food from rocks, can be exapted to take food pellets from the water surface, especially if the fish concerned has effective defences, and the food is soft.

Future genetic assimilation could easily build this kind of exapted behaviour into the genes of the plec population. It could be selected for, and then adapted along an evolutionary trajectory, so that a plec would take food from the surface *normally* in just this way.

In fact, something of the kind has probably happened already – though not in descendants of the Mathematics Institute plec, which had none. The fish in question is the upside-down catfish *Synodontis nigriventris,** which takes insects from the surface of the water in the wild using a similar technique. We have, then, both ends of a plausible evolutionary trajectory. It starts, perhaps, with a hungry catfish alerted to a food mass on the surface, near it in shallow water; perhaps a rotting, floating insect carcass. The catfish turns over in its attempts to get its mouth near to the tempting morsel, and even if it mostly makes a hash of this, any occasional success is rewarded. It will now be sensitised to this source of food, and might haunt the shallows for more of them. Its offspring, growing up in the same environment, are then more likely to be selected for similar behaviour if genetic changes can make it more effective.

This scenario contradicts Stephen Jay Gould's assertion in *The Flamingo's Smile* that adaptations like the upside-down feeding of the flamingo, scooping up crustaceans from saline lakes, must involve a single radical departure from the normal use of the beak. Animals can try out little behavioural experiments, and if they are rewarded, these can become built into their subsequent behaviour. Then, if the reward is as important as a new source of food or novel access to mates, natural selection can improve it.

Technical evolution can avoid such time-wasting, progeny-wasting, and new-function-wasting aspects of organic evolution in two ways. The first, we have discussed already: human minds can attempt to

---

* The name means 'dark belly', because when it swims upside down, its back has become light like most fishes' bellies, and its belly dark.

jump into the adjacent possible and see if it works 'in the imagi-nation'. Can we imagine an aeroplane ten times the size, and what would need to be changed for it to work? If we exaggerated the length of a bicycle frame and had the cyclist lean back, how could he see the road? Do we then want him on his front? Both have been tried, and are excellent examples: technical results of our imagination playing in the adjacent possible.

The other trick that minds can do to improve technology is to copy: to take a technical trick used in one invention and to spread its use to others. That trick, except for a few cases where genes have achieved horizontal transfer between species, is impossible for organic evolution: each lineage must invent for itself. A recent spread of this kind has been the use of digital switches in a variety of machines from toasters and children's toys to automobiles. The big one before that was the use of plastics to replace metals in the nursery, kitchen and laboratory. Before that, transparent plastics, mostly acrylics, had been used to replace glass in many applications. The progressive use of semiconductor technology is giving us solar panels, tiny refrigerating or heating elements, and a new family of very efficient light bulbs: white-light LEDs. Banks of coloured LEDs can now be tuned to give different lighting conditions; bright white light is not conducive to sleep and can be replaced with softer tones. Flexible television/computer screens, which can be rolled up like paper, already exist in the laboratory, and are not far from commer-cial production. An entire book has just been encoded in DNA, and a human face has been printed on a human hair.

In biological evolution, it used to be thought that environmen-tal 'niches', such as predatory behaviour, were already available and waiting to be occupied, rather as though some cosmic script had already written down all the possible things that an organism might do. Now it is thought that organisms construct niches as they evolve; for instance, you can't occupy the dog-flea niche until there are dogs.

Even taking copying into account, the analogous questions of competition and niche-construction in technology are as important as they are in the natural world, and they too force the evolution of new products. A good example was the colonisation of the market-place in the 1970s by VHS videotapes, even though its rival Betamax was much better in several respects. As in natural ecologies, it often happens that a less-adapted, often foreign, invader exploits the ecosystem more effectively, forcing the demise of well-established local species. The grey squirrel, for example, carries a disease that decimates indigenous red squirrels, much as the Spanish invaded South America and destroyed Inca and Maya empires. The red squirrel was better-adapted to its original environment, but the arrival of the invading grey squirrel changed that environment; in particular, it now included grey squirrels and their disease organisms. The change was sudden, biological warfare rather than the usual sedate pace of natural selection in a slowly changing environment.

In the technical world, then, there do exist processes resembling those of organic ecosystems. Many of them are recursive, affecting their own development: supermarkets make their own ecosystem of consumers, just as dogs create a new niche for dog-fleas. This makes questions about the design of technology much more difficult, because there are few real innovations, but many exaptations, copyings and adaptational trajectories. Only a few really novel tricks can be claimed to have a human designer in a non-evolutionary sense.

There is a trajectory of development for a technological product: a car starting from carriages and an engine, steam or internal-combustion; a radio starting from a crystal set and headphones; a bicycle starting as a penny-farthing and evolving through the sit-up-and-beg still seen all over China and India to the mountain bikes and lie-down versions of the latest adaptive radiation.

These are paths through our cultural history, and they make their own contexts as they evolve. The car creates vast and important areas of our cities where cars are built, where auto workers live, where the suppliers of parts have some of their factories and warehouses. When we give little Johnnie a bicycle on his seventh birthday, we introduce him to a new world that has grazed knees, gears, punctures, comparison with Fred's bike ... When the transistor radio erupted into Western culture in the 1960s, it changed the relationships of teenagers to each other and to pop stars, though nothing like as much as the mobile phone has changed all of our lives in the last few years. Alexander Graham Bell, on a promotional tour of his invention the telephone, so impressed one city's mayor that he is said to have declared: 'What a wonderful invention; every town should have one.'

Artefacts evolve, and the functions they perform get better, wider, cheaper. But they also change the society around them, so that their 'improved' next generation already has the ground prepared for it. The Ford Model T would not have been viable without gas stations, which had appeared to service the much more expensive previous generation of automobiles. In turn, the Model T and other similar affordable automobiles with privacy in the back seat changed much of the sex life of the young men and women who had access to them. Society's rules change as the Ford Model T, the transistor radio, central heating, subway travel and mobile phones affect their context, and the context in turn constrains or directs the further evolution of the product.

Nearly all inventions don't follow that kind of successful path; like nearly all species of organisms, they prosper for a little while but then die out. The few that do survive find a trajectory that takes them into the future. Frequently they move into a whole new phase space of possibilities, where their original design is effectively useless, but the new world now has an *improved* design. Like a genuine Stone-Age axe that's had its handle and blade changed several times, we find a new world with a new artefact and a new function.

In *The Science of Discworld III* we described how apparently rigid limitations on the energy needed to put a person or cargo in orbit around the Earth could, in principle, be overcome by changing the context. If you use a rocket, the amount of energy needed to get a 100-kilogramme man up to synchronous orbit can be calculated using Newton's laws of motion. It is the difference in potential energy caused by the planet's gravity well. You can't change that, so at first the limitation seems foolproof.

In the mid-1970s, however, a wholly new suggestion was made: the space bolas. Essentially this is a giant Ferris wheel in orbit. The traveller gets into the cabin as it swings past the upper atmosphere, and gets out again when it approaches the furthest point from Earth. A succession of such gadgets can deposit him in synchronous orbit a few weeks later.

A third step in the ladder of technology, not practical yet but already being discussed by engineers, is the space elevator. The science fiction writer and futurologist Arthur C. Clarke was one of several people who had this idea: take a 'rope' up to synchronous orbit and let it down to an equatorial landing-strip. The result would be a material link from a point in synchronous orbit to the ground. Once this is set up, a system of cabins and pulleys-and-weights like those used in skyscraper elevators could take a person up to orbit very efficiently. Counterweights, or another man coming down, would reduce the cost to that of the energy required to override friction.

The point is not whether we can do this yet. We can't; even carbon-fibre 'rope' is too weak. But the space elevator shows how a design trajectory can take a function away from its earliest, primitive constraints, so that a whole new set of rules applies, and the old limitations become … not invalid, but irrelevant.

More familiar examples of this 'transcendent' process are writing and telecommunication. The first attempts at writing probably involved scratches on rock or bark, and these matured in two directions – pictorial and phonetic writing. Pictorial writing, such as ancient Egyptian hieroglyphs and modern Chinese, has found it difficult to move up the technological ladder. They are not even at the rocket stage; fireworks, perhaps. Phonetic writing was more suitable for printing – the space bolas stage of the technology. This was improved as far as the great newspaper printing presses of the twentieth century and the electric typewriter. Then came the space elevator stage, word processing by computer. Ironically, this may just have saved Chinese ideograms, now easily typeset by computer, from oblivion. A further stage is starting to appear with eBooks and iPads. Eventually, all writing might be virtual, encoded in physically tiny memories until it needs only to be actualised on screens and in minds.

Communication at a distance started with semaphore and chains of watch-fires on hilltops. Navies developed coded systems of flags for communication between ships. Discworld inventors developed the clacks, a mechanical telegraph with repeaters at limit-of-sight, aided by telescopes, while we used a signal-box and mechanical link-ages to signal to trains miles from the box. With electricity came the ability to send signals via cables, and the telegraph was born. Several different coding systems for commercial transactions, and a primi-tive fax machine, were in commercial use before 1900. All these were rocket-ships. Then came the telephone, which uses sound waves to modulate an electrical signal. Much capital investment went into wiring the countryside and undersea cables to connect the conti-nents. These heroic ventures were comparable, in technical difficulty, with putting up a space bolas now. Meanwhile 'wireless' began to be used: radio, and later television. With mobile phone technology, depending upon billions of pounds of investment in immensely sophisticated base stations and in research to improve and develop the handsets, we are now beginning the space elevator stage of tele-phone technology.

We can compare these technical innovations to developments in organic evolution. We analyse the development of mammals on two scales, to show how the evolutionary process outgrows its initial constraints and achieves new properties and functions as the trajectory changes direction. We choose the two scales to emphasise that this is not a description of what actually happens during organic evolution.

We have already met the question of the extent of diversity in the animals of the Burgess Shale, and the differences of opinion between Gould and Morris. Among these animals from the Cambrian explosion, several were early chordates, ancestral to our own group of animals, including today's fishes, amphibians, reptiles, birds and mammals, as well as a diversity of modern oddities like sea squirts and lampreys. The Burgess Shale fossil *Pikaia* is the best-known early chordate, but there are others in similar Australian and Chinese fossil beds.

The early chordates produced a great adaptive radiation, firstly of jawless armoured fishes, then of a substantial number of jawed forms, including sharks, rays and bony fishes. Some of the latter, in the Devonian period, came out onto the land as early amphibians. These aquatic forms are/were the rocket-ship phase of chordate existence. The amphibians, and their diverse reptilian descendants, such as dinosaurs, birds and mammal-like reptiles that included our ancestors, constitute the next step up, the chordate space bolas. The third stage was achieved separately, and rather differently, by birds and mammals. Birds specialised in warm-bloodedness and efficient lung ventilation for flight, so that they had to provide food for their young, caring for them in nests until they could adopt the very demanding lifestyle of their parents. Mammals became turbo-powered by maintaining a stable high body temperature, and invaded many more habitats than birds, from burrowing and swimming to flying. Which they now do nearly as well as birds, but without flow-through lung ventilation. From a wide-screen chordate viewpoint, mammalian design is their space elevator.

Within that last step, we can find a similar series of invasions of the adjacent possible, in which terrestrial ecosystems were themselves changed by the presence of large land animals. Grassland such as savannah and steppes, arctic birch, lichen and moss tundra are all maintained by continuing interactions with large herbivorous mammals. Vast numbers of small rodents – mice, rats, voles, lemmings, hamsters – live in and under these grasslands. They eat more of the vegetation than their larger cousins do, and they contribute more to those ecosystems. Some interactions between mammals and their environment are familiar: rabbits making warrens, badgers excavating setts, deer ringing trees. We have to visit zoos to see the full adaptive radiation, including those strange rodents of the South American pampas: pacas, capybaras and cavies (guinea-pigs). And bats. And porpoises, dolphins, toothed whales and filter-feeding baleen whales. And all of the primates, including us. So mammals, like insects among the invertebrates, are the big terrestrial success story.

In terms of our space-exploration analogy, the mammal-like reptiles of four hundred million years ago, and today's monotremes (egg-laying oddities like the echidna and the duck-billed platypus) are the rocket-ships. The marsupial mammals – kangaroos, potoroos and opossums – are the space bolas. The placental mammals – most of today's mammals, including cows, pigs, cats, dogs, hippos, elephants, monkeys, apes and humans – are the space elevator.

Any evolutionary series can be presented as a ladder of emergent properties, new ways of being that obey new rules and have effectively discarded the old constraints. This vision is as appropriate to mammals as it is to writing tools or radio receivers. It is a general property of our self-complexifying planet in its self-complexifying universe. As time passes, more different things happen in more ways, with new rules and new functions.

That vision, of the multifarious universe knotting itself into patterns that themselves build upon previous patterns, is almost perfectly opposite to the twentieth-century view of ever-increasing entropy leading to heat death. Can this self-complication continue infinitely? We don't know, but it is as sensible a view as its opposite, and there is considerable evidence for it. Does that mean that *anything* not possible now will necessarily be possible in future? Of course not. At each step upwards, there is selection among possibilities.

This selection process is what mathematicians call symmetry-breaking: more possibilities seem to be available beforehand than are actualised at the next stage, yet paradoxically there are more possibilities afterwards than before. If advancement is the rule, and it seems to be, then contingency and selection are making up the future by evolving from rocket-ships into future space elevators, almost everywhere. We should perhaps be surprised that Moore's Law has worked for so long, but when we examine the changes in computing technology over the last decades we see that, just as in the recent mammal story, the improvements were always inconceivable at the earlier step.

This is why blinkered applications of laws of nature, such as conservation of energy or the second law of thermodynamics, can be misleading. As well as content, laws have contexts. A law of nature may appear to pose an insuperable barrier, but if you have applied the law in an inappropriate context, you may have left a way for nature to sneak round it. And it will.

FIFTEEN

CASE FOR THE PLAINTIFFS

 The Great Hall in the palace had been opened to all-comers with, of course, a podium for Lord Vetinari and desks for the lawyers. A number of guards surrounded his Lordship, and everyone heard him tell them loudly, 'No, I am in my own palace, in a court of law at the moment, and since we are not talking about a murder or a dreadful crime I see no reason to introduce weaponry into what is, when all is said and done, a philosophical debate.'

Marjorie watched the unhappy hangers-on disappear into the body of the hall, and then was further impressed by the way Lord Vetinari achieved *silence*. It was a masterclass; he simply sat silent and immobile with his hands spread out in front of him, oblivious to all the laughter, chattering, gossiping and arguing. It seemed that the air was just full of fragments of nothing whatsoever, fractured words breaking up and fading, until the final chattering fool suddenly found a great hush filling the room, in the centre of which there was his last stupid, idiotic remark, evaporating in his Lordship's dreadful, patient silence.

'Ladies and gentlemen, I cannot conceive of a more interesting case than the one we have today. The dispute is over a mere artefact: shiny, I grant you, and attractive in its way. I am given to believe by the wizards and natural scientists of Unseen University and elsewhere that, reasonably small though it be, it is in fact larger by many orders of magnitude than all of our own world.

206

'I intend to seek evidence of this during the deliberations of this very *unusual* tribunal, which has been brought into being because there are two parties who both profess to believe that the artefact is theirs. For my part, I intend to test this assumption.' Lord Vetinari sighed, and said, 'I rather fear the term "quantum" will make an appearance; but these are, after all, modern times.'

Marjorie had to put her hand over her mouth to stop giggling; his lordship had said *modern times* like a duchess finding a caterpillar in her soup.

Lord Vetinari looked around at the crowd, frowned at the desks in front of him, and said, 'Mister Slant, who is a foremost arbiter of the law, will assist me and advise me on aspects as relevant.' He raised his voice and continued, 'This, ladies and gentlemen, is not a criminal court! Indeed, I am slightly at a loss as to what kind of court it *is*, since the law works in the temporal sphere with its feet firmly on the ground. Therefore, with the two parties in this case planning to engage a number of, shall we say, experts in the celestial, as well as in the mundane sphere—' Lord Vetinari looked around and said, 'Shouldn't I have a gavel? You know, one of those things judges bang on the table. I feel quite naked without one.'

A gavel was acquired from somewhere at speed and handed to his Lordship, who banged it once or twice in a kind of happiness.

'Well, this seems to work; and now I call the counsel for the plaintiffs. Over to you, Mister Stackpole; you have the floor.'

Marjorie craned to see Mister Stackpole, but could only make out the top of a head. The voice emanating from it had a curious tone, as if its owner was actually vibrating. He said, 'A small point, my Lord, but I am a priest of the Omnian faith, and generally addressed as "Reverend".'

Lord Vetinari looked interested and said, 'Really. I shall make a note of that. Please continue, Mister Stackpole.'

Marjorie really wished she could see the face of the Reverend Stackpole. Her father, when he was alive, had quite liked being called

A second difficulty emerges from the observed 'rotation curves' of galaxies. Galaxies do not rotate like a rigid object: stars at different distances from the centre move with different speeds. Stars in the galaxy's central bulge move quite slowly; those further out are faster. However, the stars outside the central bulge all move with much the same speed. This is a puzzle for theorists, because both Newtonian and Einsteinian gravity require the stars to move more slowly in the outer reaches of the galaxy. Virtually all galaxies behave in this unexpected manner, which conflicts with numerous observations.

The third problem is the 1998 discovery that the expansion of the universe is accelerating, which is consistent with a positive non-zero cosmological constant. This was based on the High-z Supernova Search Team's observations of the redshift in Type Ia supernovae, and won the Nobel Prize for Physics in 2011.

The prevailing cosmological wisdom deals with these problems by bolting on three additional assumptions. The first is inflation, in which the entire universe expanded to a huge size in an extraordinarily short time. The figures are shocking: between 10^{-36} and 10^{-32} seconds after the Big Bang the volume of the universe multiplied by a factor of at least 10^{78}. The cause of this rapid growth – an explosion far more impressive than the wimpy Big Bang that started it all – is, we are told, an inflaton field. (Not 'inflation field': an inflaton is – well, a quantum field that causes inflation.) This theory fits many observations very well. The main snag is the absence of any direct evidence for the existence of an inflaton field.

To solve the problem of galactic rotation curves, cosmologists propose the existence of dark matter. This is a form of matter that can't be observed by the radiation it emits, because it doesn't, not in any quantity that can be observed from here. It's entirely reasonable that a lot of the matter in the universe might not be observable, but what we can infer indirectly leads to the conclusion that whatever dark matter may be, it's not made from the fundamental particles that we know about on Earth. It's a very alien form of matter, which mainly interacts with everything else through the force of gravity. No such particles have ever been observed, but there are several competing suggestions for what they ought to be, the front runner being WIMPs (weakly interacting massive particles). Despite a lot of theorising, the precise nature of dark matter is up for grabs.

The acceleration of the expansion of the universe is attributed to 'dark energy', which is little more than a name for 'stuff that makes the expansion accelerate' – though, to be fair, supplemented by detailed analyses of what kind of effect this stuff must have, and suggestions for what it might be. One possibility is Einstein's cosmological constant.

Until recently, these three *dei ex machina* resolved most significant discrepancies between the naive Big Bang theory and increasingly sophisticated observations. The introduction of these three items of novel physics, all produced out of a hat and without much independent observational support (other than what they were invented to explain), could be justified pragmatically: they worked, and nothing else seemed to. But there is now a growing realisation that the first of those statements no longer holds, but unfortunately the second still does. A growing minority of cosmologists suspect that three *dei ex machina* is at least two too many for comfort.

It is now realised that if an inflaton field exists, it doesn't conveniently switch on once and then cease to operate, which is assumed in the usual explanation of the structure of our universe. Instead, the inflaton field can swing into action anywhere, and at any time, repeatedly. This leads to a scenario called eternal inflation, with our region of the universe being just one inflated bubble in a bubble-bath of cosmic foam. A new period of inflation might start in your living room this afternoon, instantaneously blowing up your television set and the cat by a factor of 10^{78}.

Another problem is that almost all inflationary universes fail to match ours, and if you restrict initial conditions to get the ones that do, then a non-inflationary universe that performs just as well is vastly more probable. According to Roger Penrose suitable initial conditions not requiring inflation outnumber those for an inflationary universe by a factor of one googolplex – ten to the power ten to the power one hundred. So an explanation not involving inflation, although it requires an extraordinarily unlikely initial state, is massively more plausible than an explanation that does involve inflation.

A few mavericks have been devising alternatives to the standard model all along, but now mainstream cosmologists are also having to rethink the theory. There is no shortage of ideas. In some, there is no Big Bang; instead, there is a kind of revival of the steady-state universe, in which a suitably clumpy distribution of matter can survive for hundreds of billions of years, perhaps indefinitely. The redshift is not caused by expansion, but by gravity. Dark matter is not needed to explain rotation curves: instead, relativistic inertial dragging, in which rotating matter carries space along with it, might do the job.

Perhaps more radical is the proposal that either our theory of gravity, or our theory of motion, need to be modified slightly. In 2012 the particle physicist and Nobel prize-winner Martinus Veltman, when asked 'Will supersymmetry explain dark matter?', replied: 'Of course it won't. People have been looking for this stuff since the 1980s and are just talking ballyhoo. Isn't it more likely that we don't understand gravity all that well? Astrophysicists believe in Einstein's theory of gravity with a fervour that is unbelievable. Do you know how much of Einstein's theory has been tested at the distances of galaxies where we "see" dark matter? None of it.'*

The best known proposal here is MOND, Modified Newtonian Dynamics, suggested in 1983 by Mordehai Milgrom. The basic idea is that Newton's second law of motion may not be valid for very small accelerations, so that acceleration is not proportional to the force of

---

* Martinus Veltman, coming to terms with the Higgs, *Nature* **490** (2012) S10-S11.

gravity when that force is very weak. There is a tendency to assume that MOND is the only alternative to general relativity; the correct statement is that it is the most extensively explored one. Robert Caldwell,* in a special issue of a Royal Society journal devoted to cosmological tests of general relativity, wrote: 'To date, it appears entirely reasonable that the observations may be explained by new laws of gravitation.' In the same issue Ruth Durrer[†] pointed out that the evidence for dark energy is weak: 'Our single indication for the existence of dark energy comes from distance measurements and their relation to redshift.' The rest of the evidence, she says, merely establishes that distances estimated by redshift measurements are larger than those expected from the standard cosmological model. Something unexpected is going on, but it might not be dark energy.

Our confidence that we know how our universe began is being shaken. Some modified version of the Big Bang may well be correct – but then again, maybe not. When new evidence comes along, scientists change their minds.

Though perhaps not quite yet.

---

* Robert R. Caldwell, A gravitational puzzle, *Philosophical Transactions of the Royal Society of London* A (2011) **369**, 4998-5002.
† Ruth Durrer, What do we really know about dark energy? *Philosophical Transactions of the Royal Society of London* A (2011) **369**, 5102-5114.

NINETEEN

DOES GOD WIGGLE HIS FINGERS?

 Marjorie had been lost in her furious thoughts for an unmeasured length of time which, as it turned out, was about five minutes. These were broken into by Mustrum Ridcully, who had given her an appropriate nudge. She shook herself, stood up straight (which she generally did anyway) and said brightly, 'This is going to be round two; yes?'

Ponder Stibbons hurried over, detected a certain look in her eye, and said, 'Really, Miss Daw, please leave it all to the Archchancellors. After all, it is *our* business.'

Marjorie smiled: not the smile she had for a good book well read and catalogued and subsequently handed to the appropriate reader – a process she thought of as carrying *the flame.**

The chamber was buzzing as people poured in, chattering. Lord Vetinari, apparently refreshed, was ascending the stairs to the podium. The gavel dropped like thunder and, almost immediately, so did the noise.

---

* At this point it must be said that Marjorie also had a smile for a gentleman known as Jeffrey, who travelled the world inspecting, reviewing, cataloguing and pricing – and *in extremis* also restoring – the libraries of a very large number of people and organisations across the world. The two of them had an understanding, and understood quite a lot, especially about Bliss. In case anybody is now thinking of librarian pornography, this is an alternative way of cataloguing books: a system created by Henry E. Bliss (1870–1955), still in use in America and specialised libraries.

'Ladies and gentlemen, I ask the wizards of Unseen University to defend their ownership of the Round World, although it appears to me that *stewardship* might be a better and more appropriate term. It also occurs to me that I haven't even *seen* this curious thing. It is apparently reasonably small, so I will have it on my podium right now, so we can all visualise what is at the centre of today's little escapade. It will be brought to me *at once*.'

Ponder Stibbons was dispatched in haste to the university and returned, breathless, carrying the padded baize bag. Against a background of laughter, giggles and outright tittering, he gently put the contents of the bag on a tripod in front of the Patrician, who himself seemed somewhat amused by what had been placed before him.

There was a twinkle in his eye as he said, 'Excuse me if I seek for clarity, gentlemen, but could this indeed be a living world with a population of millions? Over to *you*, Archchancellor. I must say I am all agog!'

'In fact, your Lordship, I will delegate this job firstly to Ponder Stibbons, head of the Inadvisably Applied Magical facility. What he does not know about quantum – yes, I am afraid we *must* use the term, my Lord – just isn't worth knowing. Mister Stibbons—?'

Ponder cleared his throat. 'My Lord, Roundworld came into being several years ago when we were experimenting with raw firmament. The Dean experimentally put his hand in the container and wiggled it about ...'

Ponder's voice faded as he saw Lord Vetinari's expression. The Patrician was writing notes on his papers on the desk in front of him, and now he looked up, blinked, and said out loud, 'Wiggled it about? May I ask whether he intends to wiggle anything today?' A titter ran around the room as Lord Vetinari added, 'Shouldn't he be wearing gloves? I have no real ambition to be transmogrified!'

Ponder Stibbons, after he stopped laughing, rose to the occasion. 'It is unlikely, sir: we've tried and it only works with raw firmament, and *that* is very difficult to get hold of these days. If I may continue ... ?

The firmament in this case reconstituted itself as a universe, somewhat similar to our own, though happily using up only local firmament supplies. It is our belief, based *on experimentation*, that Roundworld picked up some aspects of our own world, but alas with rather less firmament. Nevertheless, it turned out to be quite small but ingeniously formed in most respects and, I might say, punching above its weight.

'We have explored other universes by various occult means, and frankly, my Lord,' he added, 'so many of them are rather drab – just a few stars banging together occasionally and with planets where there is little or no life at all. And such as it is, life there is snivelling and grovelling underground, or at the bottom of the sea, if the planet has even been lucky enough to have one of those!'

'Mister Stibbons, in your opinion, when the Dean – who I believe we shall hear from shortly – "wiggled his fingers" in firmament, did he then become a *god*?'

'Not at all, my Lord. He was nothing more than a random event, turning an instability into coherence – the same as a last snowflake just before an avalanche. Not the best way of putting it, but I think it will suffice for now. However, as a result, the whole business left certain effects in both Discworld and Roundworld; for example, Roundworld has traditions of wizards, unicorns, trolls and dwarfs; not to mention zombies, werewolves and vampires. Our research shows that although these things don't appear in Roundworld, the *concept* of them is shared by both worlds.'

Ponder took a deep breath and continued, 'The idea of gods has permeated cultures in both worlds. In our world gods are not only acknowledged but also, occasionally, *seen*. Although there are claims that they *have* been seen on Roundworld as well, the evidence is generally patchy, and sometimes simply wishful, thinking.'

'Really,' said Lord Vetinari. 'I am surprised. Gods have their uses and a part to play, and I often thank Saponaria when getting into the bath; she generally arranges the very best of suds – wonderfully fine, smooth and plentiful. Of course, I never neglect a candle for Narrativia before I embark on a lengthy memoir. It would also appear that the small gods, the household gods, survive very well. I take leave to wonder what went wrong in Roundworld?'

Marjorie's self-control finally snapped. 'Such concepts of the gods as there were on Roundworld didn't work!' she cried. 'Proud people and smart people started to put their ideas into the mouths of the gods, and shamefully it has not been unusual for two countries, ostensibly both running on the rules of the same one sacred God, to nevertheless engage one and another in combat such as never been seen on *this world* – the deliberate destruction of whole cities and even attempts to slaughter whole races. Today, many of those who saw the name of God invoked as part of this dreadful pantomime have stepped back and very much prefer reason to faith, because it is self-checking.'

Lord Vetinari sat for a moment taking this in. Then he stared at Marjorie like a cat assessing an amazing new type of mouse, and said, 'I do not believe I know your name, *madam*, or your occupation; be so good as to enlighten me, will you?'

TWENTY

DISBELIEF SYSTEM

 Roundworld has its own home-grown Omnians. We're not referring to the great majority of religious believers, who are entirely normal people who happen to have been brought up in a culture that has its own distinctive set of beliefs in things that lack objective evidence. Neither are we referring to Roundworld's equivalent of mainstream Omnians, who since the overthrow of the extremist Vorbis and his rerun of the Inquisition (see *Small Gods*) have been decent-enough sorts and kept themselves to themselves.

No, it is the Vorbises of Roundworld who cause all the trouble. Believers with a capital B. These are the people who not only *know* that their worldview is The Truth – the sole truth, the only truth, the truth revealed from the mouth of God himself – but are intent on forcing it onto everyone else, whether they want it or not, at any cost.

Most sane, rational human beings learn quite early on that you feel just as certain even when you're wrong: the strength of your belief is not a valid measure of its relation to reality. If you have scientific training, you may even learn the value of doubt. You can certainly have religious beliefs and still be a good *scientist*; you can also be a good *person* and understand that people who disagree with your beliefs need not necessarily be evil, or even misguided. After all, most of the world's people – even the religious ones – probably think your beliefs are nonsense. *They* have a different set of beliefs, which *you* think are nonsense.

But religious extremists seem unaware of the human tendency towards self-delusion, and decline to take even the simplest steps to counteract it. When the British Humanist Association hired a bus to tour the UK with the advert 'There's probably no God. Now stop worrying and enjoy your life' on its side, the immediate response from some religious authorities was: 'They don't seem terribly confident about it.' No, what they did with the 'probably' was to try not to let opponents score an easy point by criticising them for being dogmatic. Being too confident of their view. More practically, they were also worried about potentially breaching the Advertising Code. Another response from some of those of a religious persuasion was synthetic outrage and claims of persecution.

But Humanists are just as entitled to put their views on the side of a bus as tens of thousands of churches worldwide are to stick 'The wages of sin is death' on their walls. That's why the Humanists hired the bus – one small voice crying out against the multitudes, many of whom were clearly intolerant.

Belief is a very odd word, and it is used in several ways. 'Belief that' differs greatly from 'belief in', which is again different from 'belief about'. Our belief *about* science, for example, is that it's simply our best defence against believing (in) what we want to. But we may also have, to some extent, a belief *in* science, as distinct from belief in a religion or a cult: we believe *that* science can find ways out of humankind's present difficulties, ways that are not available to politics, philosophy or religion.

There is also a different usage of 'belief' altogether, one that we suspect is not always appreciated. Suppose that a scientist says 'I believe that humans evolved', and a religious person counters with 'I believe humans were created by God'. On the surface, these are similar statements, and it's easy to conclude that science is just another kind of religion. However, in religion, once you believe something, then you consider it to be an immutable truth. In science, the same word means 'I'm not very sure about this'. As we might say 'I *believe* I left my credit card in the pub', when we haven't a clue where it's gone.

Ponder Stibbons believes that Roundworld is a construction whose genesis was events on Discworld. We, and you, believe the converse: that Discworld is a construct, created by Terry Pratchett in Roundworld. It's just possible for both of these beliefs to be true – for a given value of truth. We all have beliefs of one kind or another. Let's look at how we get them, and how we might judge them.

Do newborn babies have beliefs? Surprisingly, the answer seems to be 'yes'. They are very primitive, ill-formed beliefs, and they are considerably refined even in the first six months of life, but a few behaviours, even of newborns, suggest that a lot of wiring-up of the brain has gone on in the womb. The baby is far from being a blank slate on which anything can be written – a stance that Pinker argues persuasively in his book *The Blank Slate*. The baby is especially responsive to the sight of its mother, and can become very disturbed if she simply disappears from view. It responds to music that is similar to what it heard while in the womb in the later stages of its development; it can distinguish jazz from Beethoven or folksong by attentively 'listening' for familiar sounds. It has a whole suite of beliefs about suckling, about breasts and what they're for. These things are beliefs in the sense that the baby's brain already holds some model of mother, and of music, and it prefers things that fit this model.

Soon, the baby begins to smile in response to a smile; even to a drawing of a smile. Is that a belief too? The answer depends on, but also illuminates, what we mean by a belief. The baby acts in particular ways – smiles, or suckles – because of the way its brain is wired up, because of programmes in its brain that could be otherwise, and, in occasional babies, *are* otherwise. Mostly, these are pathologies; apart from different musical preferences, there are few normal differences between baby brains. But very soon, because of a mother's behaviour, whether the baby is swaddled or carried on a bare back into the fields, or left out on a mountainside, or has its feet bound, babies *diverge*. And very soon, they are inducted into the Make-a-Human-Being kit that is characteristic of, and specific to, each human culture.

There are several ways to look at how a baby interacts with its surroundings. When the baby throws out toys from its pram, for example, this can be read in at least two ways. On the one hand, we might simply assume that it cannot retain a good grasp of the toy, which falls. However, observing the radiant smile with which it welcomes the return of the toy, we might conclude that the baby is teaching its mother to fetch. Such apparently minor interactions have a strong effect on the baby's future, and they complicate it in ways that often reinforce the culture concerned. They include little songs and stories; learning to walk, to talk, and to play. We say 'learning' here, but these processes are like birds learning to fly. Many features of the ability are already wired into the brain, but now they have to be adjusted in a kind of dialogue with the real world. 'If I stretch this bit out, and pull it back, what happens?' So these abilities *mature*: they are not learned from scratch.

In *Unweaving the Rainbow*, Dawkins likens juvenile humans to caterpillars, voracious in their uptake of information, especially from parents: Father Christmas, Heaven, fairies, what food to eat at festivals. He points out how credulous we *must* be as juveniles, to avoid obstacles to learning; but also how we should become more sceptical as adults, and that too many adults fail to do so, hence, alas, astrologers, mediums, priests and the like.

We can see just how indiscriminately juveniles pick up information through something that happened to Jack. He ran an extramural class in animal-handling for about thirty years, and became very impressed by the distribution of animal phobias (although he did realise that this was a very peculiar group of students in that respect). About a quarter of the students had a spider phobia, rather fewer had snake phobia (which, if bad, included worms). Some had a phobia for rats and mice. A few reacted badly to birds, feathers or bats. It seems likely (but we can't document it in this instance) that these phobias came about by cultural infection: Mother screamed when she found a spider in the bath, or a television series depicted snakes as poisonous. (Less than 3% actually are, but it might be wise to assume lethality as a default, for solid evolutionary reasons.) Rats are often depicted as being dirty, and the same goes for mice. Jack never worked out what gave rise to phobias about birds and feathers, but it certainly passes on in families, and it's much more likely to be learned rather than genetic. It might be a great example of how beliefs can pass from brain to brain like a computer virus, in this case not transmitted verbally. But we can see how useful these phobias would have been when we were much nearer to nature. They let us learn what creatures to avoid, instantly. And while it didn't much matter if we occasionally avoided an animal that was actually harmless, the same mistake the other way round could be disastrous.

Beliefs are formed through interactions between an individual's brain and his or her environment, especially other people but also the natural world (spiders!) So it's worth taking a general look at interactions.

If A acts on B, we call this an action; but if B also (re)acts on A, we say that A and B are *inter*acting. A baby and its mother are like that. But most interactions are not just some sort of exchange, and they have a deeper effect: A and B are, to a greater or lesser extent, *changed* by the interaction. They then become A′ and B′; then they interact again, and again, and are changed still more. After several changes of this kind, A and B have become quite different systems.

For example, the actor walks out onto the stage, and the audience reacts; the actor reacts to this, and the audience in turn reacts to the actor's new persona … and so on. In *The Collapse of Chaos* we called this deeper kind of interaction 'complicity', giving a familiar word a technical meaning that is not too far removed from the usual one, but also hinting at a mix of complexity and simplicity. The complicity between child and mother, later between child and teachers, then with sports teams, then with the whole adult world, is the Make-a-Human-Being kit we talked of earlier. We also need a word for this cultural interaction, and have suggested 'extelligence'. Individuals are *in*telligent; there are useful ideas and abilities somehow represented, remembered and readied for use, inside their brains. But most of a culture's collective knowledge is outside any given individual, forming a body of information that is not in any one brain, but *outside*; hence *ex*telligence. Before the invention of writing, most of a culture's extelligence resided in the entire collective of brains, but when writing came along, some of it – often the most important to the culture – didn't need a brain to contain it; only to extract and interpret it. Printing boosted the role of this type of extelligence, and modern technology has led to its dominance.

Where do our beliefs come from? From complicity between our intelligence and the extelligence that surrounds it. This process continues into adulthood, but its greatest effect occurs when we are children. St Francis Xavier, co-founder of the Jesuits and a mission-ary, is quoted as saying 'Give me the child until he is seven and I'll give you the man'. A trawl of today's premier extelligence, the inter-net, will haul up an almost endless range of interpretations of that phrase, from benign to malign, but their common ingredient is the malleability of human intelligence at an early age, and its fixity thereafter.

Until fairly recently, almost all people were religious believers. The majority still are, but the proportions depend on culture in a dramatic way. In the United Kingdom, about 40% say they have no religion, 30% align themselves with one but do not consider them-selves in any way religious, and only 30% say they have significant religious beliefs. An even smaller proportion attends some kind of place of worship regularly. In the United States, over 80% identify with a specific religious denomination, 40% say they attend services weekly, and 58% say that they pray most weeks. It's an intriguing difference between cultures that have such a lot in common.

Most religious activity, for the last few thousand years, is based on belief in a god or gods that acted to create the world, human beings, the beasts of the field, plants – everything. We discussed some of these creator gods in chapter 4; they used to resemble human beings or animals, but nowadays they are often abstract and ineffable; either way, they have supernatural powers. They are believed to be in daily contact with the world, making thunderstorms, providing good and bad luck for individual people, acting as a source of wisdom and authority through oral tradition (maintained by a shaman, a priest, or a priesthood). And, in the last few thousand years, Holy Books. Such theist beliefs contrast with deist beliefs, in which there is no overt anthropomorphic god, but some entity, or process, looks after the whole caboodle in deep background.

Such beliefs can be very powerful, and they form the basis of most people's views of the world and of our lives. In the seventeenth and eighteenth centuries there was a strong movement among intellectuals to reform the structure of society, by basing it on reason, rather than on faith and tradition. This movement, known as the Enlightenment or the Age of Reason, was highly influential throughout Europe and America. It played a role in the formulation of constitutional declarations of human rights, among them the American Declaration of Independence and the French Declaration of the Rights of Man.

Since then, the proportion of non-believers has increased throughout the Western world, especially among those who are well educated and well heeled financially (as a survey in the United States has clearly shown, for example). Such people, among whom we count ourselves, agree with Dawkins, though perhaps not so publicly: they maintain that there is no god, or God, out there: it's all done by laws of nature, sometimes 'transcended' by changing the context for those laws.

Good and bad 'luck' come from our own actions and the general cussedness of nature; there's no supernatural entity that consciously affects our lives.

Why do so many people believe in a god? Dennett's *Breaking the Spell* is an attempt to examine that question, for Christian fundamentalists, Islamic teachers, Buddhist monks, atheists, and others. He begins by pointing to the commonality of pre-scientific answers in groups of people: 'How do thunderstorms happen?' answered by 'It must be someone up there with a gigantic hammer' (our example, not his). Then, probably after a minimum of discussion, a name such as 'Thor' becomes agreed. Having successfully sorted out thunderstorms, in the sense that you now have an agreed answer to why they happen, other forces of nature are similarly identified and named. Soon you have a pantheon, a community of gods to blame everything on. It's very satisfying when everyone around you agrees, so the pantheon soon becomes the accepted wisdom, and few question it. In some cultures, few dare to question it, because there are penalties if you do.

J. Anderson Thomson Jr's book *Why We Believe in God(s)* devotes each chapter to a different reason for the existence of beliefs. It makes a good case for a Dennett-style system, and is persuasive enough that we'd expect aliens, if they have anything like the kind of social life we have, to have believed in god(s) during at least the early growth of their culture. The aliens would have to have had nurturing parent(s), tribes with a big alien as boss, and so on, but that's a reasonable expectation if they are extelligent.

People in all cultures grow up and acquire a set of beliefs. One way of looking at this is to call the beliefs that are inherited 'memes'. Just as 'genes' code for hereditary traits, so memes are intended to show the inheritance of individual items, rather than a whole belief system. A tune like 'Happy Birthday', a concept like Father Christmas, atom, bicycle or fairy – all are memes. A whole slew of memes that forms an interacting whole is called a memeplex, and religions are the best examples, which at various times and in various cultures have had, or still do have, many linked-up memes like 'There is Heaven and there is Hell …' and 'Unless you pray to *this* God you'll go to Hell' and 'You must teach this to your children …' and 'You must kill those who don't believe in this …' and so on. You will have some familiarity with other religions, and you will appreciate that we're not saying that *your* religion is like that. It's all the others, the mistaken ones …

We should look at a few belief systems, to see how they worked and whence they got their authority. We'll choose some relatively unfamiliar ones, where it's easier (for most of us) to set aside our own beliefs. If you're a Jewish Cathar Scientologist, skip this bit.

The Cathars were an odd group of Christians, existing from about 1100 until they were massacred around the period 1220 to 1250, initially by barons of Northern France empowered by the Pope, but then by the Inquisition. They believed that the material world was essentially evil, and that only the spiritual world was good. They deplored sex in general; indeed their bonhommes, or perfecti, wouldn't eat meat because it was the result of sexuality. Fish was all right: they didn't know about underwater sex – or plant sex, for that matter. They were totally celibate, and deplored sex even in marriage. There was a ceremony, prescribed for attainment of the perfectus state, a single sacrament, the consolamentum or consolation. It involved a brief spiritual ceremony to remove all sin from the credente, or believer, and induct them into the next higher level as a perfectus. It was commonly performed as death approached, so that the believer was not condemned. Belief in its effectiveness, however, was by no means universal.

Presumably their anti-sex views would weigh against having children, so that any such belief system would be likely to lose its adherents as time passes, but that seems not to have happened. They were remarkably successful in Languedoc, perhaps mostly through

259

conversion. In this they were the cultivated roses of religion, prop-
agated not through sex but by taking cuttings. Considering the
practices of Catholic priests, whose behaviour at that time was a
distinct contrast, it's not surprising there were many conversions.
That is probably why they had to be annihilated.

The Jews of Poland in the late Middle Ages were mostly
confined to ghettos, and restricted to a few trades including usury
– money-lending. Their beliefs were complicated. Males learned
Torah (Old Testament, Five Books of Moses) from a very young age,
and then graduated to *Talmud*, a compilation of commentaries on
the *Torah* by mostly-Babylonian rabbis. After the Bar Mitzvah ceremony at about age thirteen, which included reciting, and usually
singing, a piece from *Torah* and commenting on it, they continued
to study Jewish texts, especially the *Talmud* and the *Gemara* (additional rabbinical comments).

Boys who continued to study were frequently maintained by
general ghetto funds, such as they were (even today in Israel, boys of
Orthodox clans are allowed not to do national service). Females had
to learn to keep a kosher household, which involved a whole complex
of issues, not simply having kosher meat, but also separating milk
dishes from meat dishes, keeping separate cloths and cutlery as well
as dishes, and cleaning house, particularly for the Passover, which
required a different set of menus. The reward system was not, basically, Heaven or Hell; it was simply that doing these things led to a
good life, consonant with what God (Jehovah, but his name must not
be said) wanted for man, and to some extent woman.

In the 1550s the rules were collected into a great composition, the
Shulchan Aruch, by a Sephardic rabbi in Israel, or possibly Damascus.
They became the greatest compendium of Jewish law, especially for
the Ashkenazi communities of middle-Europe (Sephardi and Ashkenazi are two separate streams of Jewish culture). This belief system
has continued, with much evolution, to the present day. Jack's rabbi
has said that he's the best atheist in her congregation.

Scientology evolved from L. Ron Hubbard's earlier invention, Dianetics. L. Ron ('Elron') was a fairly successful science fiction author, but his entry into belief systems was distinctly more successful. Few scientists would agree with his claim that Dianetics was a science, but it sold a lot of books; he had audiences of thousands, and after the editor John W. Campbell described it in *Astounding Science Fiction* it really took off. Martin Gardner's claim that science fiction fans were very gullible seems to have been true. However, in the longer term Dianetics failed, and Hubbard produced Scientology, which has gone from strength to strength on the basis of a set of beliefs not very different from those of Dianetics.

Basically, the idea is that a set of 'engrams' is induced in people by their experiences (including when they were an embryo, before the nervous system develops). Engrams are records of bad experiences, especially very bad ones, which have to be erased for people to become clears – a step upwards on the evolutionary ladder from ordinary humans. People have souls, thetans, that have jumped from alien to alien over billions of years. The important issue for questions about belief is that this system derived from the imagination of one man, who failed to sell Dianetics. It now has tens of thousands of adherents, at least; it claims millions.

These are just three examples. Here are some others to consider, since people seem to pick up sets of beliefs terribly easily.

Rosicrucians, for instance, believe that a set of mystical instructions will enable them to achieve telepathy, success in their jobs and instantaneous travel anywhere, including other planets. The cost of this instruction is considerable, but eventually it gets you into the central core of the sect, where anything is possible. Atlanteans believe that every so often the Earth tilts, flooding all the present continents and exposing new ones; if you find an Atlantean, note where he buys his next house. There are hundreds of such belief systems, and the people who subscribe to them – often paying large sums of money –

get all kinds of benefits, especially being privy to the real truth about life, the universe, and everything.

Other belief systems are not so wild. We have in mind systems like Count Alfred Korzybski's general semantics, which produced wise little gems like 'the map is not the territory', Ludwig von Bertalanffy's general system theory, and the many systems of mind training such as Esalen, with which Gregory Bateson was associated. There are thousands of 'mind training' hits on Google, most of them based in California. It is easy to understand the feelings, the beliefs, that send people into these systems of self-improvement. We subscribe to some ourselves – devotion to explanations involving 'complexity', promoted by the Santa Fe Institute for Complex Systems and the New England Complex Systems Institute (whose acronym, NECSI, has enabled Jack to promote himself as a necsialist, if not quite a nexialist*).

However, the variety of these beliefs – most of which seem very strange to outsiders – is amazing. How can so many belief systems, differing so radically from the common experience of humanity, be accepted by so many people? For each individual belief system, the majority of us consider at least some of the beliefs to be absurd. So why is the absurdity not apparent to everyone? Can it be that people in general are so ignorant of reality nowadays that they will buy into anything that promises a better or more interesting life?

There was also a system advertised not that long ago which forecast that 2012 would be a year of financial collapse *and* the beginning of World War III – which wouldn't of itself have been a great surprise given some of the conflicts. However, the forecast was based on rather strange reasoning: not as a result of the antics of greedy bankers and the armaments industry, but because the ancient Mayan calendar ran out in 2012.† The Mayans themselves mostly ran out in

* Science fiction author A.E. Van Vogt coined the term in *Voyage of the Space Beagle*. He defined a nexialist to be someone who is good at joining together, in an orderly fashion, the knowledge of several fields of learning.

† It didn't, anyway. The period concerned was just the first of an even vaster series of calendar cycles.

the 1600s, because of the diseases which the Spaniards brought, not because of Spanish military prowess. So it's difficult to see what their calendar had to do with us. The calendars on many kitchen walls this year – and most years – run out on 31 December ... Hallelujah! It's the apocalypse!

In 2012 *Scientific American** reported a psychological study carried out by Will Gervais and Ara Norenzayan, under the title 'How critical thinkers lose their faith in God'. It was a follow-up to a 2011 investigation by Harvard researchers, who concluded that what we believe is closely linked to how we usually think. Intuitive thinkers, who come to conclusions instinctively, tend to have religious beliefs. Analytic thinkers tend not to. Encouraging people to use intuition rather than logical analysis increased their belief in God.

Gervais and Norenzayan wondered whether the underlying distinction could be understood in a slightly different manner, as a difference between two ways of thinking that are both useful in suitable circumstances. System 1 thinking is 'quick and dirty', relying on simple rules of thumb to make decisions rapidly. If an early human on the savannah spots a patch of orange behind a bush, it makes good sense to assume that it might be a lion, and take avoiding action. A more analytical System 2 assessment might subsequently discover that the orange patch was a bunch of dried leaves, but the processes involved would be slower, and involve more work. In this case, System 1 thinking does little harm if it later turns out to be mistaken, but System 2 could kill you if there really is a lion and you waste time trying to decide.

On the other hand, there are many occasions on which System 2 saves lives, but System 1 does not. Thinking about past forest fires, and deciding not to build your village in an area surrounded by dry

* Daisy Grewal, How critical thinkers lose their faith in God, *Scientific American* **307** No. 1 (July 2012) 26.

vegetation, trumps an intuitive assessment that the location has ample building materials. Avoiding floodplains, even though it is easy to build houses on them and they are currently unoccupied, can prevent complete destruction of your property when the river rises. There is a *reason* why they are currently unoccupied.

Teasing out the workings of the human brain is tricky, but psychologists have developed techniques that help. In this case, participants were first interviewed to determine the extent of their religious beliefs. Sometime later, the main experiment was carried out, in two different ways. In the first, participants were given a randomly rearranged five-word phrase – such as 'speak than louder words actions' – and were asked to rearrange the words to make sense. Some of them were given scrambled phrases containing many words related to analytical thinking; the rest were not. After this exercise, they were asked whether they agreed that God exists. The group whose training period involved words related to analytical thinking were more likely to disagree. Moreover, this tendency remained, even when their prior beliefs were taken into account. The second version of the experiment relied on previous research, showing that asking people to read something printed in a hard-to-read font promoted analytical thinking, perhaps because they have to proceed more slowly and puzzle out the meaning of the letters. Subjects that completed a survey printed in a semi-illegible font were less likely to agree that God exists than those given the same material in a legible one.

The magazine article summed up the study: 'It may help to explain why the vast majority of Americans tend to believe in God. Because System 2 thinking requires effort, most of us tend to rely on System 1 thinking processes whenever possible.'

There is a loose relationship between System 1/System 2 and Benford's distinction between human-centred or universe-centred thinking. Intuitive thinking mainly takes a human-scale view of the world, and often places emphasis on quick decisions based on little more than hunches. Many people, finding it difficult to weigh up

Here's an analogy. If you take a car, and change any single aspect even a little bit, the odds are that the car will no longer work. Change the size of the nuts *just a little*, and they don't fit the bolts and the car falls apart. Change the fuel *just a little*, and the engine doesn't fire and the car won't start. But this does not mean that only one size of nut or bolt is possible in a working car, or only one type of fuel. It tells us that when you change one feature, it has knock-on effects on the others, and those must also change. So parochial issues about what happens to little bits and pieces of our own universe when some constant is changed by a very small amount and the rest are left fixed are not terribly relevant to the question of that universe's suitability for life.

Some additional sloppy thinking parlays this fundamental blunder into a gross misrepresentation of what the calculations concerned actually show. Suppose, for the sake of argument, that each of the thirty parameters has to be individually fine-tuned so that the probability of a randomly chosen parameter being in the right range is 1/10. Change any parameter (alone) by more than that, and life becomes impossible. It is then argued that the probability of all thirty parameters being in the right range is 1/10 raised to the power 30. This is 10^{-30}, one part in a nonillion (ten billion billion billion). It is so ridiculously small that there is absolutely no serious prospect of it happening by chance. This calculation is the origin of the 'knife edge' image.

It is also complete nonsense.

It's like starting at Centrepoint, in the middle of London, and going a few metres westwards along New Oxford Street, a few metres northwards up Tottenham Court Road, and imagining you've covered the whole of London. You haven't even explored a few metres in a north-westerly direction, let alone anything further away. Mathematically, what is being explored by each change to a *single* parameter is a tiny interval along an axis in parameter space. When you multiply the associated probabilities together, you are exploring a tiny box whose sides correspond to the changes made to individual parameters – without considering changing any of the others. The car example shows how silly this type of calculation is.

Even using the constants for *this* universe, we can't deduce the structure of something as apparently simple as a helium atom from the laws of physics, let alone a bacterium or a human being. Our understanding of everything more complex than hydrogen relies on clever approximations, refined by comparison with actual observations. But when we start thinking about other universes, we don't have any observations to compare with; we must rely on the mathematical consequences of our equations. For anything interesting, even helium, we can't do the sums. So we take short cuts, and rule out particular structures, such as stars or atoms, on various debatable grounds.

However, what such calculations actually rule out (even when they're correct) are stars just like those in this universe and atoms just like those in this universe. Which isn't quite the point when we're discussing a different universe. What other structures could exist? Could they be complex enough to constitute a form of life? The mathematics of complex systems shows that simple rules can lead to astonishingly complex behaviour. Such systems typically behave in many different interesting ways, but not in just *one* interesting way. They don't just sit there being dull and boring, except for one special 'finely tuned' set of constants where all hell breaks loose.

Stenger gives an instructive example of the fallacy of varying parameters one at a time. He works with just two: nuclear efficiency and the fine structure constant.

Nuclear efficiency is the fraction of the mass of a helium atom that is greater than the combined masses of two protons and two neutrons. This is important because the helium nucleus consists of just that combination. Add two electrons, and you're done. In our universe, this parameter has the value 0.007. It can be interpreted as how sticky the glue that holds the nucleus together is, so its value affects whether helium (and other small atoms like hydrogen and deuterium) can exist. Without any of these atoms, stars could not be powered by nuclear fusion, so this is a vital parameter for life. Calculations that vary only this parameter, keeping all others fixed, show that it has to lie between 0.006 and 0.008 for fusion-powered stars to be feasible. If it is less than 0.006, deuterium's two positively charged protons can push each other apart despite the glue. If it is more than 0.008, protons stick together, so there would be no free protons. Since a free proton is the nucleus of hydrogen, that means no hydrogen.

The fine structure constant determines the strength of electromagnetic forces. Its value in our universe is 0.007. Similar calculations show that it has to lie in the range from 0.006 to 0.008. (It seems to be coincidence that these values are essentially the same as those for nuclear efficiency. They're not *exactly* equal.)

Does this mean that in any universe with fusion-powered stars, both the nuclear efficiency and the fine structure constant must lie in the range from 0.006 to 0.008? Not at all. Changes to the fine structure constant can compensate for the changes to the nuclear efficiency. If their ratio is approximately 1, that is, if they have similar values, then the required atoms can exist and are stable. We can make the nuclear efficiency much larger, well outside the tiny range from 0.006 to 0.008, provided we also make the fine structure constant larger. The same goes if we make one of them much smaller.

With more than two constants, this effect becomes more pronounced, not less. Numerous examples are analysed at length in Stenger's book. You can compensate for a change to several constants by making suitable changes to several others. It's just like the car example. Changing any one feature of a car, even by a small amount, stops it working – but the mistake is to change just that one feature. There are thousands of makes of car, all different. When the engineers change the size of the nuts, they also change the size of the bolts. When they change the diameter of the wheel, they use a different tyre.

Cars are not finely tuned to a single design, and neither are universes.

Of course, the equations for universes might run contrary to everything that mathematicians have ever seen before. If anyone believes that, we've got a lot of money tied up in an offshore bank and we'd be delighted to share it with them if they will just send us their credit card details and PIN. But there are more specific reasons to think that the equations for universes are entirely normal in this respect.

About twenty years ago, Stenger wrote some computer software, which he called MonkeyGod. It lets you choose a few fundamental constants and discover what the resulting universe is capable of. Simulations show that combinations of parameters that would in principle permit life forms not too different from our own are extremely common, and there is absolutely no evidence that fine-tuning is needed. The values of fundamental constants do *not* have to agree with those in our current universe to one part in 10^{30}. In fact, they can differ by one part in ten without having any significant effect on the universe's suitability for life.

More recently, Fred Adams wrote a paper for the *Journal of Cosmology and Astroparticle Physics* in 2008, which focuses on a more limited version of the question.* He worked with just three constants – those that are particularly significant for the formation of stars: the gravitational constant, the fine structure constant, and a constant that governs nuclear reaction rates. The others, far from requiring fine-tuning, are irrelevant to star formation.

Adams defines 'star' to mean a self-gravitating object that is stable, long-lived, and generates energy by nuclear reactions. His calculations

* Fred C. Adams, Stars in other universes: stellar structure with different fundamental constants, *Journal of Cosmology and Astroparticle Physics* 8 (2008) 010. doi:10.1088/1475-7516/2008/08/010. arXiv:0807.3697.

reveal no sign of fine-tuning. Instead, stars exist for a huge range of constants. Choosing these 'at random', in the sense usually employed in fine-tuning arguments, the probability of getting a universe that can make stars is about 25%. It seems reasonable to allow more exotic objects to be treated as 'stars' too, such as black holes generating energy by quantum processes, and dark matter stars that get their energy by annihilating matter. The figure then increases to around 50%.

As far as stars go, our universe is not improbably balanced on an incredibly fine knife edge, battling odds of billions to one against. It just called 'heads', and the cosmic coin happened to land that way up.

Stars are only part of the process that equips a universe with intelligent life forms, and Adams intends to look at other aspects, notably planet formation. It seems likely that the results will be similar, debunking the almost infinitesimal chances alleged by advocates of fine-tuning, and replacing them by something that might actually *happen*.

What, then, went wrong with the fine-tuning arguments? Failures of imagination and blinkered interpretations. For the sake of argument, let us accept that most values of the constants make atoms unstable. Does this prove that 'matter' cannot exist? No, it just proves that matter identical to that in our universe can't exist. What counts is *what would happen instead*, but advocates of fine-tuning ignore this vital question.

We can ask the same question for the belief that the only viable aliens will be just like us, as many astrobiologists still maintain – though fewer of them than there used to be. The word 'astrobiology' is a compound of astronomy and biology, and what it mostly does is put the two sciences together and see how they affect each other. To analyse the possibility of alien life, especially intelligent alien life, conventional astrobiology starts with the existence of humans, as the pinnacle of life on Earth. Then it places them in the context of the rest of biology: genes, DNA, carbon. It then examines our evolutionary history, and that of our planet, to find environmental features that helped bring life, and us, into existence.

The upshot is an ever-growing catalogue of special features of our, and Earth's, history, alleged to be necessary for alien life to exist. We mentioned some of these features earlier; now we'll discuss some of them in more detail. They include the following conditions. Life needs an oxygen atmosphere. It needs water in liquid form. That implies being at a suitable distance from the Sun – the much-emphasised habitable or Goldilocks zone, where temperatures are 'just right'. Our unusually large Moon stabilises the Earth's axis, which would otherwise change its tilt chaotically. Jupiter helps protect us from comet impacts – remember how it sucked up Shoemaker-Levy 9? The Sun is neither too big nor too small, both of which make terrestrial planets less likely. Its rather dull and boring position in the galaxy – not at its centre, but out in the boondocks – is actually the best place to be. And so on and so on and so on. As the list grows ever longer, it is hard not to conclude that life is extraordinarily unlikely.

An alternative approach, which we like to call xenoscience, reverses the direction of thought. What are the possible types of habitat? We now know, as we did not until recently, that there is no shortage of planets. Astronomers have found over 850 exoplanets – planets outside our solar system – enough to provide a statistical sample that suggests that there are at least as many planets in the galaxy as stars. The physical conditions on those planets vary enormously, but that provides new opportunities for new kinds of life. So instead of asking, 'Is it like Earth?' we should ask, 'Could some form of life evolve here?'

We're not even restricted to planets: subsurface oceans on moons whose surfaces are thick layers of ice would be a good place for life, even for Earthlike life. We should take into account local conditions, but we should not assume that features that appear favourable in our solar system necessarily apply elsewhere. Without a large moon, a planet's axis may indeed tilt chaotically, but it could do so on a scale of tens of millions of years. Evolution can cope with that; it might even be enhanced by that. Life in a big enough ocean wouldn't even notice. A large gas giant may sweep up comets, but that could slow evolution down, because the occasional catastrophe adds variability. Jupiter may keep comets at bay, but it greatly increases the number of asteroid impacts on the Earth. The current best estimate suggests that Jupiter has done more harm than good, with regard to life. Some life forms such as tardigrades (commonly called waterbears or moss piglets) resist radiation better than most of those on our planet. The rest don't need to, because the Van Allen belts, regions of electrically charged particles maintained by the Earth's magnetic field, keep radiation away. In any case, if the belts hadn't been there, life could have become more tardigrade-like.

The so-called habitable zone is not the only region around a star where life might be possible. Some exotic chemical systems can make life-like complexity possible without water, and liquid water *can* exist outside the habitable zone. For example, if a world close to its star is tidally locked, so that one side perpetually faces the star and the other faces away, there will be a ring-shaped twilight zone on the boundary between the two faces, where liquid water might exist. Worlds far from the star can have liquid oceans underneath an outer coating of ice: Jupiter's moon Europa is the best-known example in the solar system, and it is thought to have an underground ocean containing as much water as all of Earth's oceans put together. The same goes for Ganymede, Callisto and Saturn's moon Enceladus. Titan – another moon around Saturn – has liquid hydrocarbon lakes and an excess of methane, hinting at non-equilibrium chemistry, a possible sign of unorthodox life.

The idea of a galactic habitable zone – the claim that alien life can exist only in the region of the galaxy with enough heavy elements but not too much radiation – is especially controversial. The Danish astronomer Lars Buchhave and his team have surveyed the chemical composition of 150 stars, with 226 known planets smaller than Neptune. The results show that 'small planets ... form around stars with a wide range of heavy metal content, including stars with only 25 per cent of the sun's metallicity'. So an excess of heavy elements is *not* required for Earthlike planets. NASA scientist Natalie Batalha remarked that 'Nature is opportunistic and prolific, finding pathways we might otherwise have thought difficult'.

And so on and so on and so on.

Life adapts to its environment, rather than the other way round. Goldilocks doesn't have the final word: Daddy Bear and Mummy Bear have valid opinions too. What is 'just right' for life depends on what kind of life. So-called extremophiles exist on Earth at temperatures below freezing and above boiling. It's a silly name. To such creatures, their environment is entirely comfortable; it is *we* who are extreme. It's even sillier to use the same name for creatures in two environments so different that each creature would consider the other to be even more extreme than us.

The second approach is far more sensible: instead of successively cutting down the opportunities for life, it explores the full range of the possible. That vast and impressive shopping list of features 'necessary' for life, making humans seem extremely special, is poor logic. Life on Earth demonstrates that the list is *sufficient* – but that doesn't make it *necessary*.

These two ways of thinking about aliens are of course yet another example of Benford's dichotomy. Astrobiology is human-centred, because it starts from us and narrows the universe down until it fits. Xenoscience is universe-centred: it keeps possibilities as broad as possible and sees where they lead. We are beautifully adapted for our environment because we evolved to be like that. This observation is much more reasonable than claiming that we humans are so special that the solar system, the galaxy, even the entire universe, was constructed in order to accommodate *us*.

Cosmic balance …

Is life really balanced on a knife edge, then? Or have we got it all wrong?

Let us go back to our rod and sharp knife experiment. It seems undeniable. Try again to balance the rod on the cutting edge of the knife. However carefully you place it, it tips and slides to the floor. There is no question: the balance has to be extraordinarily precise.

The mathematics is, if anything, even more compelling. The masses on each side, multiplied by their distances from the knife, must be *equal.* Exactly. The slightest imbalance leads to total failure. So, by analogy, any imbalance in the laws of nature, however insignificant, would destroy the conditions required for life to exist. Change the speed of light or various other constants by a few per cent, and the delicate carbon resonance in stars would fail. No resonance, no carbon, no carbon-based life.

Maybe, though, we've accepted these arguments too readily. How relevant, how sensible, is the analogy of a metal rod and a sharp knife? Straight metal rods are an artificial product of technology. In mathematics and nature, most things are nonlinear – bent. What happens if you place a bent rod on top of a knife edge? Assume the bend is not too great, and roughly in the middle. Provided you place the rod on the knife so that it's reasonably near the balance point, as soon as you let the rod go it turns so that the free ends hang downwards. It slips sideways, but not very far, and then it stops. For a few seconds it seesaws up and down, but eventually it comes to rest.

Perfectly balanced.

Reach out a fingertip and push one end up a little. When you let go, the bent rod swings back to its original position, overshoots, reverses direction, and eventually settles back to where it was to begin with. If you push the other end down, the same thing happens.

Next, move the rod sideways on its pivot, away from the bend. The shiny metal is slippery, and the rod slides back until it balances again. It's not necessary to *arrange* for the rod to balance. It does so of its own accord. At the balance point, the forces pulling it to either side cancel out just as precisely as they would have to do to balance a straight rod, but the rod no longer falls off if the balance is wrong. It just moves a little, and finds its own balance point. The mathematical reason is straightforward. The rod seeks a state of minimum energy, where its centre of mass is lowest. Because the centre of mass of a bent rod is below the pivot, it ends up hanging in a stable position.

It's not *necessary* to fine-tune the universe.

It can fine-tune itself.

The 'knife edge' thought-experiment is rigged; the analogy with nature is false. The experiment depends on the rod being *straight*. Pretty much any other shape would be self-correcting. In fact, even a straight rod will balance on your finger. As long as the finger is close to the midpoint, the rod no longer slides off. Agreed, a finger is sweaty and sticky, and that can stop the rod sliding, but that's not the main reason why the rod balances. If one end tilts upwards, the rod rolls sideways and the point of contact with the finger moves away from the raised end. The weight of rod on the raised side is now greater than that on the other side, so the combined forces conspire to return the rod to the horizontal. If it is tilted the other way, the same thing happens. Even a straight rod will find its own balance point if the pivot is not a knife-sharp edge.

Not only is the thought-experiment rigged: so is the metaphor. A universe doesn't have to be perfectly linear, and it doesn't have to pivot on an infinitely thin line. The anthropic, human-centred mentality has unerringly homed in on exactly the wrong metaphor. It ignores the universe's tendency to respond to change by altering its own behaviour.

The triple-alpha reaction in the red giant star is just like that. An exact coincidence of energy levels is not necessary. The nuclear energy of beryllium plus that of helium is within a few per cent of one of the energy levels of carbon – but not spot-on. That's where the red giant comes in. The energies balance only if the star is at the right temperature. *And it is.* This may seem to be even further evidence of fine-tuning: the astrophysics of the red giant has to compensate precisely for the disparity in nuclear energy levels. But the star is like the bent rod. It has a nuclear thermostat. If its temperature is too low, the reaction proceeds faster, and the star heats up until the energies become equal. If the temperature is too high, the reaction proceeds more slowly, and the star cools down until the same thing happens. It would be just as sensible to admire the exquisite precision with which a wood-burning fire adjusts its temperature to be exactly that at which wood can burn. Or to be amazed that a puddle fits exactly into the dip in the ground that contains it.

The knife edge analogy depends on linear thinking – that's why it uses a straight rod. But we live in a nonlinear universe, in which anything that is stable automatically tunes itself so that it works. That's what stability *means.*

Natural systems are like your arm, not like the knife. This is how the triple-alpha process tunes itself so exquisitely, and why your legs are exactly long enough to reach the ground. It is also why we, as evolved creatures, are so neatly adapted to the universe we inhabit. Analogous beings living in different universes would *also* be exquisitely adapted to their local conditions. This is why most of the Goldilocks arguments, that life elsewhere in the universe must be just like it is here, are probably nonsense.* There are many genuine mysteries here, much to marvel at, and much yet to be understood. But there is no compelling scientific reason to believe that the universe was specially made for us.

We are faced with two alternatives. Either the universe was set up in order to bring us into being, or we evolved to fit it. The first is human-centred: it raises humanity above the universe in all its awe-inspiring vastness and complexity. The second, a universe-centred

---

* See Jack Cohen and Ian Stewart, *What Does a Martian Look Like?*

view, puts us firmly in our place: we are perhaps an interesting development, complicated enough that we don't understand exactly how it all works, but hardly the be-all and end-all of existence.

We have been around for a few million years at most, perhaps only 200,000 if you restrict attention to 'modern' humans; the universe is about 13.5 billion years old. We occupy one world orbiting one of 200 billion stars in one galaxy, which itself is one of 200 billion galaxies. Isn't it just a tiny bit arrogant to insist that the entire universe is merely a by-product of a process whose true purpose was to bring us into existence?

TWENTY-THREE

OVER-ZEALOUS ZEALOT

 Afterwards, Marjorie reflected on that aspect of the afternoon.

Roundworld was the planet Earth, in theory anyway, and surely flinging it around and sloshing it about would cause the seas to also slosh about a bit, to say the least. Nevertheless, she automatically fielded the globe, which against all reason slotted into her palm with a decisive but moderate stinging sensation, which had gone in a second.

The hooded man glared at her and drew out a curved knife. She could see the play of light on the blade and wondered how good her unarmed combat skills would prove to be against an opponent who clearly knew how to use a knife, especially since she was almost totally out of breath. The man screamed, 'Om is good!' and swung the blade at her.

Marjorie jumped backwards and a very large wolf landed in between, just as a hailstorm of bats dropped out of the sky. For a moment Marjorie stared fixedly at this tableau, and then, well, it all got quite exciting. Suddenly the wolf had the knife and the man was on the ground, and the bats had disappeared in a flurry to be replaced by a naked young woman, who looked both ways along the alleyway and said, 'Excellent work for a civilian! You ought to get a medal!'

Still clutching Roundworld like a hot water-bottle, Marjorie managed to say, 'But look! There's still a wolf!'

The wolf stood up on its hind legs, and the girl said, 'Better turn your head. Captain Angua does not like to be seen when she is – how can I put this? – well, deshabillée. Give her some space, please.'

Against all reason, Marjorie turned her back on the wolf, listened for a few seconds to what sounded like an autopsy in reverse, complete with unpleasant gurgling noises, and then a new voice said, 'I'm impressed. Some people throw up just by listening. Allow me another minute to get into this dress, and we'll be right with you.'

Indeed, only a couple of seconds later she realised that she really *was* in the company of a couple of young women – both now dressed – who showed her what looked very much like police badges. She recognised them as policemen anyway – she sometimes had to get them to call in at the library if one of the usual suspects was acting up, and policemen always looked rather out of place in the presence of literature. These two, however, seemed a whole lot smarter than the general run of the constabulary.

They cheerfully told her that they were indeed a vampire and a werewolf, the vampire introducing herself as Captain Sally and the wolf-lady as Captain Angua, before adding with a grin, 'But don't you worry, miss, we don't eat on duty.'

In Marjorie's bemused state it all seemed perfectly normal as the three of them then waited until a wagon turned up and disembogued them of their over-zealous zealot.

'I believe Lord Vetinari would like a word, miss,' the policewoman who had been a wolf then said.

'What? I distinctly heard the hooded people say they were going to kill him!'

Angua shook her head, and said, 'People *try* occasionally; sometimes he lets them live – even with all their bits if they are entertaining enough: he has what they call a mercurial sense of fun. On this occasion I can report that the group of zealous Omnians who attacked him were defenestrated.' Angua smiled and added, 'You have to hand it to his Lordship. He has style and is remarkably stronger than you might think. Lord Vetinari jumped out of the window and *refenestrated* them back into the hall!'

It was two days later when Marjorie Daw once again dined in the hall of Unseen University. In the centre of the feast, Roundworld glittered and shimmered miraculously, as became a world that could be in two places and be two different sizes all at the same time.

There were, of course, toasts and more food than was good for anybody. Lord Vetinari, who was also there, said, 'I believe, madam, that you could stay if you wish, but I understand that you have declared to the Archchancellor that you want to go back to … let me see … oh yes, the library of the borough of Four Farthings, England, wherever that may be. Are you sure?'

Marjorie smiled and said, 'Oh yes, very sure; there is no telling what the council will do if I'm not there. Probably halve the budget and fill the place with anodyne Good Citizenship displays and other idiocies. Politicians only read books they have written, or those of colleagues they suspect might have mentioned them in their text. Or they simply want to pretend that they have read the latest touted bestseller to show that they are just like "the common people", neglecting the fact that people aren't *all* common and can spot a phoney at a glance.' She paused, then added, 'Sorry for the rant, sir, but I just had to get it out of my system. I've got to get back before they replace me with a yahoo who doesn't even know where the damn word came from.'

She let Lord Vetinari refill her glass, and felt a lot better.

The following lunchtime, on the lawn in Unseen University, the Great Big Thing hung in the air, scintillating, twisting, coruscating, evaporating and gently spinning. It was, in a very strange way, alive, and yet not alive: like people are alive, and ships are alive, or even mountains – in their own strange way – are alive, but alive all the way through.

Surrounding it was the usual squash of fervent young white-robed wizards, muttering about 'thaumic energies' and 'slood derivatives' and the kind of terminology that made Rincewind's head ache. Their fingers were almost *twitching* in their eagerness to get going

305

on the next stage of the Great Big Glitch … oops, no, the Great Big Experiment.

Ponder Stibbons was also there, with other members of the Inadvisably Applied Magic group, and of course all the senior wizards, who would not miss something like this, even for lunch. Ponder, after all the hand-shaking, said, 'Well, Marjorie, I'm sure we are all sorry that you have to go, but I only need to press the button in front of me to put you back just where you were before you so abruptly ended up on this turf. As the Archchancellor said, it is doubtful that we will do this *particular* experiment again. Sometimes even wizards know when *not* to meddle.'

In the silence that followed, a high-pitched excitable voice could be heard from amongst the crowd of young wizards: 'You know, I think I know what we got wrong …'

Just then the Librarian of Unseen University knuckled his way across the turf at speed. He stopped when he reached Marjorie, blew her a kiss and handed her a banana.

She blew back the kisss with an extra *s* as Ponder said, 'I have looked for a suitable sentence to speed you on your way, Marjorie, and came across a much-liked one: *What goes around comes around.* Welcome to Roundworld! It's only a page away.' Then he pressed the button. 'So, you will be back home before I have finished this senten—'

TWENTY-FOUR

NOT COLLECTING STAMPS

Although it is widely held that faith can move mountains, it has not reliably been seen doing so. Yes, of course it's a metaphor – a powerful one, and a valid one. People have done, and will continue to do, amazing things because of their beliefs. But the main things that move mountains significantly are subducting tectonic plates, volcanic eruptions and earthquakes. Oh, and rain and cold, given long enough.

There is no denying the power that faith has over human beings, and the sometimes remarkable acts that it can motivate, but it really is a curious way for *Homo sapiens* to behave. It requires acceptance of a rather strange mixture of moral precepts and the supernatural. There is no direct objective evidence for many beliefs that are central to the world's great religions – but there are innumerable reports of miraculous events, holy people, longstanding authority and rituals that may go back thousands of years. Religions are grounded in deep culture, inculcating the present generation's values in the next. And they are often desirable values, don't get us wrong.

However, there is an evident danger if you ground your morality in authority and ineffable deities. What is moral simply becomes what is prescribed. God is good – but this can lead to the concept that *anything* can be deemed good if you can convince people that God so wills. Such as cutting off the head of an infidel, or blowing women

and children to smithereens in order to get yourself into Heaven –
typical tactics of Roundworld's own over-zealous zealots. With a few
exceptions of that kind, largely to do with who counts as a genuine
person, most of the world's religions have their prized moral values
in common. However, they are little more than the standard default
values of most human societies. Don't kill people. Don't steal. Don't
do anything that you wouldn't like done to you. Nearly all of us can
sign up to these values, be we Christians, Jews, Muslims, Hindus, Jedi
Knights … even agnostics and atheists. It is not necessary to invoke a
god to provide 'authority' for them. They are the common currency
of humanity.

That leaves the supernatural elements for us to disagree about, and
that's where the real trouble starts. Those elements matter, because
they endow a religion with its cultural significance. Anyone can sign
up to 'don't kill people', but only we Righteously Reformed Rince-
windian Roundworldists genuinely *believe* that the entire universe is
a foot across and sits on a shelf in Unseen University.

Prove us wrong.

We're sitting in the audience, and there's a debate in progress on the
stage. The protagonist is very sure of his position, has good clear
pictures, and is very clear about his story. His antagonist is different.
She is rather unsure; her pictures are sketches and cartoons, and she
is altogether more tentative.

Which do we tend to believe?

It mostly depends on who *we* are.

There are some who like certainty; they like to know just where
they are. They tend to get their knowledge, their beliefs, from author-
itative sources: the Bible, the Quran, textbooks, or the practices of
their professions. They *know* that those who disagree with them are
at least wrong, and sometimes evil. It's certainly more than sinful
for politicians to change their position on almost any topic. They
simply can't understand why someone can't see the Truth when it's

presented to them, or that someone can't appreciate the clarity of their assertions or the power of their arguments.

Over the years we have found, somewhat to our surprise, that many scientists are also like this. In private, they often acknowledge that there are difficulties with the current state-of-the-art theories in their subject area. They may even accept that some key features might have to be changed as more evidence comes in. But their public face is one of complete certainty. There are biologists who *know* that the most important feature of any organism is its DNA, and that virtually everything about living creatures is explained by their genes. There are physicists who *know* that the universe is made up of *these* particles, with *these* constants and mechanisms. They know that, ultimately, everything in the world reduces to fundamental physics. We can see that engineers can very easily adopt this position about their subject; after all, it is almost entirely man-made: gears, engines, oscilloscopes, MRI machines, LEDs, cyclotrons … But electrons? Quantum waves? W and Z particles? The Higgs boson?

Others are suspicious of such certainty, tending to say 'I don't know' quite a lot, and are unsure about lots of things. Their beliefs have come from a medley of sources, many of them quite unreliable; they tend to change their minds, even about quite important issues.

Dennett's *Breaking the Spell: Religion as a Natural Phenomenon* initially takes us back to the times when people didn't have access to information of any reliable kind. But like so many New-Agers today, they took 'information' from astrology, from myths, from gossip, from folklore – because there wasn't anywhere else to get it. Extelligence, the information outside heads, was then very disorganised; but primitive religions were an exception. They were often extensively organised, with lots of gods and goddesses, a cosmology or three, ceremonies and rituals.

Religions, in fact, were the most organised ways to run your life. As time passed, some kind of natural selection among religions went on, so that the ones that survived, the ones that gained adherents, became more effective for gaining even more. The Ten Commandments was a very good set, ensuring that there were less social problems even if most were 'More honor'd in the breach than the observance'. 'Eat rotting meat' would have been a bad one. 'Love your neighbour' was remarkably good (initially in Judaism, then in Christianity), then spreading through the next 1500 years, according to a suggestion in Pinker's *The Better Angels of our Nature* about the universal decline of human violence.

Now that extelligence has become better organised, with such things as internet search engines to help us navigate through overwhelming quantities of information, we can look back and see the beginnings of rationality among the Egyptians and the Greeks; then to some extent among the Romans and the Hebrews; then the Reformation and the Enlightenment. Rationality, and the beginnings of science, Bacon and Descartes, began to take over from theology as a way to run life, at least for a few people – those who wrote the tracts, anyway. From steam-power and canals and trains, via the industrial revolution, this led to the modern world.

However, religions remained as a backdrop to the play. Priests were always there to give their blessings, or to curse advances in rationality. Galileo, persecuted by the Church for his belief that the Earth went round the Sun, stands for thousands of such episodes. The Catholic Church has recently admitted it was in the wrong on that occasion, though rather grudgingly, and with growing ambivalence. But what about all the others, minor and major?

Among Western people, a solid proportion are now basically rational in their approach to life and its problems, but about 30% run their lives in strict accordance with religious tenets of one kind or another. Nothing like that many regularly attend churches or synagogues, but most Muslims go to mosques. The majority don't give the way they should live a lot of thought; they run their daily lives as a matter of habit, conditioned by whim … Is that really too pessimistic a statement? How many people get home from work, turn the television on and their minds off?

Mobile phones and the internet are helping, but the attitude to these is often closer to religion than rational: they are seen as supernatural, worked by demons, perhaps. You know what we mean, if you come from the era before mobile phones: they're miraculous. As Arthur C. Clarke wrote: 'Sufficiently advanced technology is indistinguishable from magic.' This was the main theme of *The Science of Discworld*, especially in Benford's alternative form 'Technology distinguishable from magic is insufficiently advanced'.

Many Cambodians, especially those in the hill tribes, are animists. They believe that spirits are everywhere: in the water, the trees, the clouds. They have shamans, tribal 'doctors'. In 2011, Ian gained an interesting insight into shamans when visiting a Cambodian village. A child was ill, and the shaman was performing a ceremony to expel bad spirits and restore her health. The interesting part was that the tribe had sent her to a conventional doctor the day before, who had put her on a course of antibiotics. Naturally, the shaman had to ratify this with the right ceremony, thereby making it possible to take the credit. The villagers presumably saw little difference between the antibiotics and the ritual – but someone in the tribe, perhaps the headman or one of his two wives, had the sense to try both. Human- and universe-centred thinking in an unholy alliance.

The world's major religions dismiss animism on the grounds that belief in several gods – polytheism – is ridiculous. The intelligent way to go is monotheism, belief in one god. (Or, in the case of Unitarianism, belief in *at most* one god.) But is monotheism the great step forward that is so unquestioningly assumed?

It has a definite attraction: unification. It assigns all of the universe's puzzling features to a single cause. Belief in one god is less off-putting than belief in dozens. It's even consistent with Occam's razor.

If you want to invoke Thomas Aquinas's ontological argument for the existence of God, in his *Summa Theologica*, monotheism is unavoidable. There, he invites us to consider 'the greatest conceivable being'. If it did not exist, then there would have to be a greater conceivable being: one that did exist. That surely is greater than a non-existent greatest being. So God exists, QED. Moreover, He is unique: you can't have two greatest beings. Each would have to be greater than the other.

Logicians and mathematicians are painfully aware, however, that this argument is flawed. Before you can use a characterisation of some entity to deduce its properties, you have to provide *independent* proof that such an entity exists.

The classic example is a proof that the largest whole number is 1. Consider the largest whole number. Its square is at least as big, so it must equal its square. The only whole numbers like that are 0 and 1, of which 1 is larger. QED. Except, 1 is clearly not the largest whole number. For instance, 2 is bigger.

Oops.

What's wrong? The proof assumes that there *is* a largest whole number. If it exists, everything else is correct, and it has to be 1. But since that makes no sense, the proof must be wrong, and that implies that it doesn't exist.

So, in order to use the ontological argument to infer the existence of the greatest conceivable being, we must first establish that such a being exists, *without* simply referring to the definition. So what the argument proves is 'If God exists, then God exists'.

Congratulations.

At any rate, whatever advantages monotheism may possess, being a consequence of the ontological argument is not one of them.

Monotheism's supposed great triumph, unification, may actually be its greatest flaw. Assigning all puzzling phenomena to the same causes is a standard philosophical error, the equation of unknowns. Asimov put it this way: if you don't understand UFOs, telepathy or ghosts, then UFOs must be piloted by telepathic ghosts. This way of thinking invents a label and attaches it to all mysteries, closing them off in the same way. It claims the same cause for all of them, which robs that cause of any explanatory force.

If you are a Cambodian animist, believing in a spirit for every natural phenomenon, you are aware that different phenomena may have different explanations. What explains water is not the same as what explains a tree. This can be a starting point for finding out more. But if you are a monotheist, offering the same explanation of *everything* you don't understand – whatever it is, and equally applicable even if it were totally different – then you are just closing down lines of enquiry, advancing the same facile answer to every mystery.

How many people, in today's scientific and technical world, have beliefs that are consonant with the kind of world they live in? How many understand about microwave ovens, why aeroplanes can stay up, about how electricity is distributed to houses (and don't expect electricity from unconnected sockets in their wall), and how milk comes from cows, not from supermarkets? What proportion of people do we need to be rational, to keep civilisation running? More to the point, these days: how many people does it take – gangsters or terrorists, bigots or zealots – to break down the workings of a civilised society? And why should (some) religions foster that kind of terrorism, aiming to do just that? It may just be extremists, but there are clearly belief systems that encourage such extremism.

There's an answer, but we would be happier if it were wrong. People live their lives, and are acquainted with all kinds of events, but for most people it's a small world. In an African tribe, there may be fasts and festivals, intimate relationships with about twenty people, mostly relatives, and a nodding acquaintance with about another hundred; just like Orthodox Jews in Golders Green, or Muslims in Bradford. Workmates, hobbyists, football supporters, pub acquaintances and friends can bring the total up to about 150. Humans seem to be able to remember about 200 faces, at most.

In consequence, the lives of all these folk are nearly all parochial, much as life is portrayed in TV soaps. The events that happen to them are mostly small. Births, marriages and deaths are rare, coronations much rarer. It is not surprising that religions, bringing order into that narrow kind of life, setting it in a much bigger frame, are popular. They provide prayer, hymns and sermons to make such lives feel more meaningful. They promise bigger things: gods, angels and life after death. Tabloid newspapers' obsession with celebrities, people everyone has seen on TV, similarly gives ordinary lives some glamour.

But there is another, darker side. Religions that preach damnation, or that predict an imminent end of everything in some kind of cataclysm, will also be attractive because what they are concerned with is imminent, now, tomorrow, happening to me and to the people I know. Relatives and friends will be damned, or caught up in the cataclysm. We must save them! Whether they want it or not.

Religion is human-centred. Though it pretends to be universe-centred, that universe is the tiny one created by their god, whether it be Odin or Jehovah or Brahma. Like the universe of *Star Trek*, it's minuscule compared to the real thing. It is a human-sized village with its own headman, blown up to cosmic proportions but not greatly changed.

Astrology, like many other 'personal' new-age philosophies, picks up on the same attraction: what matters is what happens to *me*. Such lifestyles don't even pay religious dues (maintaining the church roof, the vicar's salary, hush-money to erstwhile children assaulted by priests or celebrities). They are belief systems that pretend to knowledge of the future, *my* future – convincingly enough to have caught more than one American president – while taking no responsibility for the accuracy of those predictions. Religions whose compass includes heaven-or-damnation contrive equally to promise and threaten without any guarantee of a blissful, or terrible, afterlife. But it's an afterlife for *me* that's at stake; deeply personal, not a bit universal. No guarantee is needed if you have faith.

Contrast that with the scientific stance. It's surprisingly difficult to find science that matters, to *me*, that isn't embodied in technology. The numbers are meaningless; even that important Sun is about 150 million kilometres away; solar storms may disrupt electronics, but not (mostly) *my* electronics. There are billions of stars in the Milky Way, billions of galaxies each like our own – but what does that do for me? There are hundreds of chemicals in our foodstuffs, hundreds of kinds of plant – mostly weeds, whose particulars are not necessary for nearly everyone – in our forests and meadows. There are millions of transistors in a computer, a mobile phone or a television. But *I* don't need to know about that to operate them; just turn them on, play games on the computer, watch EastEnders on telly. Watch nature programmes, watch science programmes. Don't get involved, as there's nothing there that seems to affect *me* directly. It's all universe-related, not people-related; it's Benford's contrast again.

A story about Jack is relevant here. When he was about fourteen, he was breeding tropical fish to accumulate money for going to university. His father had been killed dumping ammunition after the end of World War II, and his mother was earning about £2 a week as a machinist: not enough to pay rent (she had only a half-pension). Jack found a mated pair of angelfish, very rare at that time, and bought them for £50. That was a lot of money: he had about £75 in the bank, from breeding other fish. Within a week, one angelfish had died. He then bought another one, for £15.

His grandfather, with whom they were living, said (and he remembers this very vividly, especially his grandfather's 'study': one corner of the living room with piles of newspapers): 'This is where we tell if you are a queen bee or a wasp.' His grandfather didn't know much biology, and Jack remembered that un-biological aspect of the remark all these years. But his grandfather did know the distinction between having global concerns or only immediate concerns, and that's the distinction he was making.

The angelfish bred, and Jack sold the first brood for £50; they bred again six weeks later, and again and again. He made a lot of money from them. The important distinction stayed with him: he became a scientist. He gave up on becoming a rabbi, which his father had intended for himself, an intention that fell on Jack's shoulders, being the only boy. He could perhaps have taken on a pet shop, but that was not to his taste. Without understanding his grandfather's distinction – he only understood it, to his shame, when writing this chapter – he was a queen bee with global concerns, not a wasp concerned only with human-centred things.

One irony of the story: Jack had thought that the fish that had died was a male, and replaced it with what he thought was another male. It turned out that both were females; the one he'd thought was female, which survived, was actually male. Even if you are a queen bee, you still need a bit of luck. Now, it becomes clear that Jack's grandfather was asking whether Jack was human-centred or universe-centred: an Omnian fundamentalist, or a wizard.

Is a science-versus-religion argument going on now? Like there was, after Darwin published *The Origin of Species*? To read the newspapers, you could easily think that scientists are up in arms, trying to destroy religions.

Without doubt, there is a desperate anti-Darwinism prejudice in the middle states of the USA, in Indonesia, and in a few other countries. This seems to have its origin in politics rather than anti-rationality, since many of its proponents, such as those promoting the hypothesis of intelligent design, claim to be putting forward a rational, scientific criticism of Darwinism. The political aim in the USA is to get round the constitutional separation of church and state, by putting religion into the schools wrapped in science's clothing. (That's not solely our view: it's what Judge John Jones concluded when presiding over *Kitzmiller v. Dover Area School District*, when he ruled that the teaching of intelligent design in school science classes was unconstitutional.) The methodology is to present an anti-Darwin stance in schools, perhaps in order to deny 'naturalism', the belief that nature can work perfectly well without gods. Alvin Plantinga and Dennett discuss this point in *Science and Religion; Are They Compatible?* This is yet another example of Benford's distinction. Believers in, and promoters of, an intelligent designer want a human-centred system of the world. They want evolution to be guided. They have completely missed Darwin's point, that a creator is unnecessary: natural selection can produce the same results without there being any human-type design.

This anti-Darwin prejudice, this wish for a human kind of design in evolution, must be distinguished from all those places in the world that haven't yet emerged from a medieval dependence on religion in people's daily lives, and where evolution isn't 'believed in'. And it must also be distinguished from an unthinking commitment to religion, hence disbelief in evolution – or in science in general – in the lives of most people even in scientific/technological societies today.

Dennett and Thomson explain the commitment to religion very well. It is irrational and faith-based, but for many people it seems almost to be a necessary part of being human. It provides a sense of identity and a shared culture. Part of the reason is that most religions have, in the course of their evolution, changed to become more and more adapted, more appropriate to the creatures they're serving. All of their organisation, and most of their practices, have been developed better to serve their practitioners. Those that didn't do so well have been lost to history. Few people now believe in Odin or Osiris.

Modern religions, with their beliefs in gods or at least in the supernatural, have all achieved congregations that seem happy with the hierarchy of senior people who determine the letters of the faith. This complicity between congregants and the hierarchy makes the belief system almost irrelevant, even though it seems to the congregants

317

to be central. The joint activities, the singing and the praying, the individual commitments in common, give the congregants a warm feeling of belonging. From outside, each of these faiths seems a beautiful harmony, the odd spat over homosexuals or female bishops aside. It's not surprising that rationality can't edge its way in.

For decades, psychologists have been making scientific studies of religious belief; not with a view to proving or disproving the existence of any particular flavour of deity, but trying to find out what goes on inside the minds of believers. Some have concluded that belief in the supernatural is a more or less inevitable consequence of evolutionary survival value (an ironic finding, if true), because it knits human cultures together. Only recently has it occurred to a few psychologists that perhaps the thought processes of atheists also need to be investigated, since such people form a fairly large group that seems to be immune to these supposed evolutionary pressures. Comparing believers with non-believers is likely to shed more light on both.

Even if religion and other kinds of belief in the supernatural really are natural consequences of humanity's past history, built into our thought processes by evolution, there is no compulsion to continue to think that way. Our sporadic tendency towards violence, especially against each other, can also be explained in similar terms, but there seems to be a widespread (and sensible) view that this does not excuse violent behaviour. A true human being should be able to override such innate urges by an act of will. The same can be said of belief in the supernatural: by exercising our intelligence we can train ourselves to disbelieve claims for which there is no clear evidence. Of course, believers think that there *is* evidence – certainly enough to convince them – but it tends to be obscure and heavily dependent on interpretation.

An instructive example of the influence of religious belief on rational judgement occurred in 2012 when Sanal Edamaruku, founder of Rationalist International and President of the Indian Rationalist Association, was invited to examine a miracle. What follows is based on an interview with Edamaruku published in *New Scientist*, and we report what was alleged there.*

The miracle occurred at a Catholic church in Mumbai, where water was dripping spontaneously from the feet of a statue of Christ on the cross. This event was interpreted as a sign from God – a holy miracle – and flocks of believers collected and drank the water, apparently thinking that it was holy water that would cure all manner of illnesses. A television station asked Edamaruku to comment, and consonant with his position, he rejected the claim of a miracle. Since his view was at that moment purely a matter of opinion, the TV company challenged him to provide scientific proof, which of course required visiting the church and taking a look.

The church authorities gave their approval. It didn't take long to find the cause of the 'miracle'. A drainage channel from a washroom passed beneath the cross's concrete plinth. A quick look at the drain revealed that it was blocked. The walls behind the cross, and the wooden cross itself, were soaking up drainage water through capillary action. Some of the water was emerging through a nail hole and running down over the statue's feet. Edamaruku took photographs to document the cause.

Point made, you will imagine. Well, yes – but. Edamaruku had long been a thorn in the side of religious groups, and his finding caused them some embarrassment. They could have used System 2 thinking to investigate the likely causes of dripping water, or just called a plumber like most sensible people would have done when they found water dripping from places where water ought not to be. Instead, they made a System 1 judgement and plumped for a supernatural explanation. But it's not a great idea to have people drinking

* One minute with Sanal Edamaruku, *New Scientist* (30 June 2012) 27. See also http://en.wikipedia.org/wiki/Sanal_Edamaruku.

dilute sewage, even if they do imagine it's a miracle cure. The discovery probably saved the church a great deal of potential trouble, even if it debunked the miracle.

So what was the response?

The church itself did nothing. But according to Edamaruku, people from two lay Catholic associations filed charges against him under section 295A of India's penal code, which dates to 1860 and forbids 'deliberately hurting religious feelings and attempting malicious acts intended to outrage the religious sentiments of any class or community'. Edamaruku has said that he is willing to appear in court, where he is convinced the case will be thrown out – but unfortunately the law has a nasty sting in its tail. Anyone accused can be jailed, perhaps for many months, before the case comes to trial. So, as we write, Edamaruku has fled to Finland, and the Rationalist Association has set up an online petition calling for the complaints to be dropped.

Christian theologians have long worried about the paradox of *silentio dei*, the silence of God: if God exists, why does He not speak? An omnipotent, omnipresent being should have no difficulty in making His existence evident, in undeniable ways. Lined up alongside this strange absence are other problems of human existence: why a caring God permits diseases and natural disasters, for example. Theology being what it is, innumerable answers have been proposed.

There's a Jewish joke about this. (There's a Jewish joke about everything.) Three rabbis are arguing a point in theology. Two claim it was first made by Rabbi ben Avraham; the third claims it was Rabbi ben Yitzchak. 'Look, I know it was him! I studied this for my thesis!' But the others still disagree. Eventually, in desperation, the third rabbi says, 'I know, let's ask God!' So the three of them pray, and suddenly the sky splits open and God leans out, looks down, and says, 'He is right. It was Rabbi ben Yitzchak.'

After a stunned pause, the first rabbi says: 'Well, now it's two against two.'

Upon reflection, the joke works because we know it wouldn't be like that. God could solve the problem of disbelief by writing his name across the sky in letters of fire a kilometre high. But for obscure theological reasons, an omnipotent being apparently declines to exercise that particular power. The only possibility that theologians have not contemplated is that God is silent because He doesn't exist. On that particular issue all religious factions agree – and they don't accept *that* explanation.

So, if you were to take a vote, there would be a clear majority verdict: God does exist. Atheists are a definite minority. However, even if you think that questions about the universe can be decided democratically, you have to ask the question sensibly. Religious people are happy to align themselves with all of the other religions in the world when it comes to those dreadful atheists – infidels, literally people without faith. But as soon as you start to examine what different religions, or different sects within a given religion, or even different believers within the *same* sect, actually believe, common cause gives way to bedlam. The Church of England, for example, is currently split into factions over the issue of women bishops, and is perilously close to splitting into two different sects. And the Church of England itself originated in a split from the Church of Rome. There are thousands of different Christian denominations, let alone other faiths.

In this debate, we have no desire to argue for either position. We'd rather there were no bishops at all – men or women – though being realists we don't expect that to happen. What intrigues us is that good – indeed, devout and committed – Christians, people on both sides of the argument, have examined their innermost hearts, prayed to their God and been answered with a clear vision of God's wishes. There can be no doubt that that is what they sincerely believe. But, curiously, God's wishes turn out to be that (a) Women bishops should be allowed, and (b) They shouldn't. Indeed, God's wishes are remarkably similar to what those of the individuals concerned have been all along, before they consulted their deity on the matter.

From within that debate, if it can be dignified with the word, it is clear to all that one side is right and the other is wrong; one has correctly divined God's wishes, the other is deluded. Problem: which is which? From outside, we are observing an interesting experimental test of the efficacy of prayer, indeed of the existence of the kind of deity in which the Church of England believes, indeed the general concept of a belief system. *Silentio dei* is not the difficulty: God has indeed spoken to both sides – or so they genuinely believe. But He has spoken with a forked tongue. From outside, if He existed in a form consistent with the beliefs of the Church of England, then surely He would have told everyone the same thing.

So this particular religion fails a definitive experimental test, one inadvertently set up by the believers themselves. In science, that would be a good reason to reject the hypothesis.

Worldwide, religious believers outnumber atheists, even if we exclude people who nominally belong to a religion but don't practise it. However, across the board, the world's religions find it virtually impossible to agree on the supernatural features of their belief systems. They often seem to agree on fundamentals such as a god – but which god? Each religion, each sect, has a god that – it tells us – demands a different set of rituals, a different form of worship, different prayers. Each is in the minority, so at most one can be correct. But they all appeal to the *same* reasoning: faith. Since their own beliefs disagree, faith clearly doesn't hack it. Thus the apparent majority turns out to be smoke and mirrors.

The writer and comedian Ricky Gervais* made a similar point more pithily in 2010:

The dictionary definition of God is 'a supernatural creator and overseer of the universe'. Included in this definition are all

---

* http://blogs.wsj.com/speakeasy/2010/12/19/a-holiday-message-from-ricky-gervais-why-im-an-atheist

deities, goddesses and supernatural beings. Since the beginning of recorded history, which is defined by the invention of writing by the Sumerians around 6000 years ago, historians have catalogued over 3,700 supernatural beings, of which 2,870 can be considered deities. So next time someone tells me they believe in God, I'll say 'Oh, which one? Zeus? Hades? Jupiter? Mars? Odin? Thor? Krishna? Vishnu? Ra … ?' If they say, 'Just God. I only believe in the one God,' I'll point out that they are nearly as atheistic as me. I don't believe in 2,870 gods, and they don't believe in 2,869.

Ultimately, religious beliefs are based not on objective evidence, but on faith. Religions are belief systems, and many proclaim this as an advantage: faith is a test, set by God. If you don't agree with them, you've failed. Many religionists – and a proportion of postmodernists – have claimed that science is also a belief system; in effect, just an alternative religion. Not so. They have failed to understand the key difference between science and belief: in science, the highest points are given to those who *disprove* the tenets of the alleged faith, especially its central tenets. In science there is no continuing central dogma, such a strong characteristic of religions. Indeed, that is what defines any particular religion: its central creed. Rationality, or indeed science, continually matches ideas against each other – and for science, to the extent that it's possible, against events in the real world – and is prepared to change its stance according to the way they do or do not agree. For religions, in contrast, events in the real world are held up to the dogma. If they match, they are accepted; if they don't, they are either ignored or declared to be evil, needing to be destroyed.

Science can't disprove religious beliefs. *Nothing* can. That's the problem. It's like trying to prove that our universe does not sit on a shelf in Unseen University, a region of the multiverse that is forever inaccessible to us. But the inability of science to disprove religious beliefs in the supernatural does not make it a belief system, even if it may sometimes lead people not to believe in the supernatural. When presented with extraordinary hypotheses, disbelief is not the opposite of belief. It is the default, neutral stance: 'I'm not interested in playing this game, it makes no sense.'

Many religious people try to reject atheism by portraying it as merely another form of belief, with the natural position being what they call agnosticism. They then interpret that stance as the view that the chances of God existing are about 50-50. So by being neutral, you are already halfway towards agreeing with them. This is nonsense. As Christopher Hitchens has said: if we are asked to accept a proposition without evidence, we are also entitled to dismiss it without evidence.

The *default* is to disbelieve. An atheist is not someone who believes that God doesn't exist. It is someone who doesn't believe that God *does* exist. If you think those are the same, ponder this statement by the comedian Penn Jillette: 'Atheism is a religion like not collecting stamps is a hobby.'

EPILOGUE

L-SPACE

 What Marjorie Daw was not was the kind of person who goes around saying to herself, 'Oh, it must have been a dream.' But by the fourth day she was beginning to question her own sanity with some force.

Her first days home had been a whirlwind as she really got back to grips with her work: seeing to the new books, beefing up the science fiction section, arguing with the council treasurers for more funding – even with the Libraries Committee itself – and demanding that she should be the arbiter of all that came into the library and how it was displayed.

And *that* meant putting the Bible onto the fantasy shelf.

The committee looked into her eyes, and didn't disagree.

One evening, when Marjorie, who was as always the last to leave, was turning out the lights in the library, feeling rather angry because somebody had defaced a book by Richard Dawkins with a lot of squiggles and phrases like 'God is not mocked!' she thought she heard a noise and smelled a mildly pungent smell.

She found, suddenly prominent on her desk, a large ripe banana ...

Overhead, a voice said: 'Ook!'